Sustainable Development and Earthcare

Searching for Policy Initiatives

ABOUT THE EDITORS

Dr. K.V. Sundaram is a well-known Geographer-Planner, who worked with the Indian Planning Commission and various prestigious regional and local-level planning projects of the UN in India and other Asian and African countries. His publications include 16 books and more than 160 research papers. His technical expertise covers urban and regional planning, rural area development, multilevel planning, eco-development and human resource development. Presently, Dr. Sundaram is the Founder-Chairman of Bhoovigyan Vikas Foundation. He was conferred the Lifetime Excellence Achievement Award by the Centenarian Trust, in Chennai in 2004.

Dr. Mrityunjay Mohan Jha is associated with SMEC India, Gurgaon. He received his Masters in Geography, M.Phil and Ph.D. in Food Security from JNU, New Delhi, besides receiving training in Geographic Information System from the Indian Institute of Remote Sensing, Dehradun. He has been part of research/consultative group for Department of Science and Technology, Ministry of Rural Development, Ministry of Human Resource Development, Government of India, IFPRI South Asia, ADRF and World Bank. He received an Excellence Award for 'people centred work' from the Centre for Land Use Management, Hyderabad in 2005. He has to his credit four books and about a dozen research papers published in journals of repute.

Prof. Prem Shankar Tiwari obtained his Ph.D. in Urban Geography from Banaras Hindu University in 1969 and got advanced training in Regional Planning at the Settlement Study Center, Rehovot, Israel in 1979. He joined the teaching profession at Azamgarh in 1956 and later on moved to Bilaspur (Chhattisgarh), Pratapnagar, Mysore and finally to the University of Madras in 1976. He also had a short stay at Osmania University, Hyderabad. Also, he has worked as Editor, Indian Geographical Journal; General Secretary, Indian Geographical Society; Editor, Annals of the Association of Geographers, India (NAGI); and as President, NAGI. His research contributions have been in the fields of Agricultural/Rural Development, Regional Planning, Natural Resource Management and Population Issues.

Sustainable Development and Earthcare

Searching for Policy Initiatives

Edited by
K. V. Sundaram,
Mrityunjay M. Jha,
P. S. Tiwari

CONCEPT PUBLISHING COMPANY, NEW DELHI-110059

ISBN-13: 978-81-8069-652-7 ISBN-10: 81-8069-652-9

First Published 2009

Published and Printed by

Concept Since 1974

Concept Publishing Company
Regd. Office:
A/15-16, Commercial Block, Mohan Garden
New Delhi-110059 (India)
Phones: 25351460, 25351794, *Fax*: 091-11-25357109
Email: publishing@conceptpub.com,
Website: www.conceptpub.com

Editorial Office:
H-13, Bali Nagar, New Delhi-110 015, India.

Cataloging in Publication Data-- *Courtesy:* D.K. Agencies (P) Ltd. <docinfo@dkagencies.com>

Sustainable development and earthcare : searching for policy initiatives / edited by K.V. Sundaram, Mrityunjay M. Jha, P.S. Tiwari.
p. cm.
Contributed articles in Indian context.
Includes bibliographical references. Includes index.
ISBN 13: 9788180696527 ISBN 10: 8180696529

1. Environmental policy--India. 2. Nature conservation--India. 3. Sustainable development--India. 4. Human ecology--India. 5. Sustainable agriculture--India. 6. Water-supply--India--Management. I. Sundaram, K. V. (Kavasseri Vanchi), 1929- II. Jha, Mrityunjay Mohan. III. Tiwari, P. S. (Prem Shankar)

DDC 363.700954 22

Dr. M. Velayutham
Former Director, National Bureau of Soil Survey and Land use Planning (ICAR)
Former Director, M.S. Swaminathan Research Foundation, Chennai
Chairman, Bhoovigyan Vikas Foundation (Earth Care Foundation)

Foreword

The major focus of activities of the Bhoovigyan Vikas Foundation (BVF) is to work towards promoting 'sustainable development' and achieving the common goal of 'Saving Our Planet Earth'. Considering the fact that the United Nations General Assembly declared the International Year of Planet Earth, as a 3-year event (2007-2009) to increase awareness of the importance of Earth sciences for the achievement of sustainable development, the publication of this volume by BVF is apt and timely. The historic UN Earth Summit of 1992 at Rio on 'Environment and Development' signalled the importance of sustainable development, emphasizing the need to recognize and realize in our pursuits, environmental and ecological protection and economic development, as two sides of the same coin. This would call for a deeply held commitment and practice to a new ethics viz., 'the ethic for Earthcare, Sustainable Development, Sustainable Consumption and life styles', by the society at large. Mahatma Gandhi's remarks that "Nature provides for everyone's needs but not for his greeds" exhorts the practice of this Ethics.

Distinguished experts have dealt with in detail in 20 Chapters arranged in four sections in this Volume the contours of some priority policy and development issues in the natural resources domain emerging from Earth-Man interface.

The central piece of this volume is the paper by K.V. Sundaram and M. Moni—"The case for an Earthcare Policy Institute in India" with its vision and mission being to sensitise the society, the beauty and bounty of the natural resources i.e. land, ocean, atmosphere and biodiversity, and to the realization that "we have not inherited them from our forefathers but borrowed them from our grand children". In spreading this message, this volume is indeed a significant contribution and frigger.

M. Velayutham

Chennai
18.1.2009

M. Velayutham

Preface

Ever since the first UN Conference on the 'Human Environment' held at Stockholm in 1972, environment and development issues have moved to centre stage in political and public debate. The UN Conference on Environment and Development in 1992, the World Summit on Sustainable Development in 2002 and the Second International Conference on "Sustainable Agriculture, Water Resources Development and Earthcare Policies" held in New Delhi in 2002, have furthered the discussion, in search for a better earth for human beings. This book has emanated from the Second International Conference of the Bhoovigyan Vikas Foundation and the subsequent discussions with the experts on the earthcare theme.

Nature and mankind are inseparable part of life that includes land, water, air, space, energy, flora and fauna, which are all interconnected and interdependent. Man is one among the many species competing with others for survival. In the name of progress, human race has changed from hunter-gatherers to a self-centred modern man, an exploiter of nature. So much so, development in a sense has become synonymous with environmental degradation and progress with pollution. The term 'sustainable development' has acquired significance in both developed and developing countries. Sustainable development, so eminently articulated by the World Commission on Environment and Development in 1987 through the Brundtland Report titled "Our Common Future", has now become the Mantra for the management of natural resources. With the world population expected to reach 8 billion by 2030, pressure on environment will continue to mount. The major challenges of the coming years include alleviation of poverty and hunger, providing livelihood opportunities to billions, managing water resources with greater public participation and producing enough food to meet the growing requirements, among others. The land base per person is shrinking and common property resources on which the rural poor survive, are highly degraded and the protective cover of forests is declining, leading to the loss of valuable diversity.

The present century has been recognized as an "Earthcare and Knowledge Century", and many national and global issues will come to head during this period, calling for a reappraisal of our development priorities, as well as taking some bold initiatives in the policy and planning realms. All these would call for a deeply-held commitment to a new ethic, the ethic for Earthcare, sustainable development, sustainable consumption and life styles. How to translate these principles into practice by working through the systems already in place in the country, is our major dilemma. In India, land degradation and change in the land use patterns pose major problems. In urban areas, land pollution is caused by disposal of solid waste and discharge of effluents. Though we have crossed the food self-sufficiency mark, the annual rate of food grain production is on the declining trend. In addition to this, major areas of concern will be water availability for agriculture, low productivity, inability to sustain growth, ecological stress and environmental concerns. Besides, growing energy requirements pose tremendous challenges to the environment, government and society. Considering the diverse range of issues that encompass earthcare and sustainable development, his book concentrates on major issues in four different sections, viz., Earthcare Policy; Man and Environment: Sustainability Issues; Sustainable Livelihoods and Agriculture; and Issues in Water Resources Development.

Section I: Earthcare Policy

Earthcare implies a development paradigm, wherein issues in environment and economics are circumscribed by the requirements of human development and ecological sustainability. It is an overarching concept. It includes besides conservation of earth's vitality and diversity, human actions to minimize the depletion of non-renewable resources, keeping development actions within the carrying capacity of the earth. It is a multi-dimensional and multi-disciplinary concept, requiring an integrated and coordinated approach by governments, in which effective institutional mechanisms will have a crucial role. While there are institutions to take care of issues pertaining to human development, there are none to take a holistic and integrated view on the conservation and strengthening of earth's vitality, on which all life on planet depend. As most of the important issues to be addressed in this context are strongly interlinked, our actions must be mutually supportive, synergistic and aimed at the common goal. This section contains seven papers dealing with critical earthcare issues, requiring urgent attention.

K.V. Sundaram and M. Moni in "*The Case for an Earthcare Policy Institute in India*" advocate for filling up an organizational vacuum in the country. The new century is the "Earthcare and Knowledge Century", and many national and global issues will come to a head during this period, calling for a reappraisal of our development priorities, as well as taking some bold initiatives in the policy and planning realms. All these would call for a deeply-held commitment to a new ethic, the ethic for Earthcare, sustainable development, sustainable consumption and life styles. How to translate these principles into practice by working through the systems already in place in the country, is our major dilemma. This paper is an attempt to outline a proposal to fulfil the development policy requirements of our government, within the framework of a strategy for caring for the Earth and achieving sustainable development. Its acceptance and implementation would go a long way in ushering in a new era of development without destruction.

In their paper on "*Population and Environment: Facing the Challenges*", *K.V. Sundaram and Mrityunjay M. Jha* join the sustainability debate mainly in the context of developing countries. The developing countries of the world today are in a deep-seated and multi-faceted development crisis, mainly characterized by three major challenges, i.e., (i) *population explosion*, (ii) *environmental degradation* and (iii) *economic stagnation*, which is a consequence of the above two. While the economic growth indicators showed progress in these countries, the ecological indicators tell a different story, by every ecological yardstick. Today, at the beginning of the 21st century, we commence our journey on the planet, with a dubious legacy, i.e. with a heavily depleted natural capital base and a highly deficit 'ecological balance sheet'. The environmental problems have been examined in the context of population and technology. The authors have recommended action along the triple bottom line, i.e. *Economic, Social and Ecological*, which is now referred to by planners as "inclusive development".

K.V. Sundaram, in his paper on "*Policies for our Plundered Planet: Some Pathways to the Future*", seeks to outline the contours of some priority policies, and the problems emerging from Earth-Man interface and their various consequences, which call for urgent action along several fronts. The paper underlines the fact that humanity's stake in earth protection is enormous during the current century, as the sane earth care policies, necessary in this behalf, have been neglected too often in the past. The basic dilemma here is related to the two-way relationship between development and environment. This discussion is organised around the areas warranting urgent attention, viz. international environmental concerns, the crisis of multilateralism, environmental degradation, sustainable water security, sustainable agriculture and food security, sustainable energy and industry, sustainable rural development policy, population and carrying capacity and human settlements planning and environmental laws and economic instruments for promoting sustainable development.

Rekha Singhal and Nidhi Sharma have discussed the central role of natural resources management and adaptive learning in achieving sustainable development, in their paper on Integrating Traditional and Scientific Knowledge: *An Approach to Earth Care*. Natural resource management plays central role in determining sustainable development of any country. It is a subject that has gained intense public and political attention both locally and globally requiring practitioners to adapt to changing social and environmental expectations. This has transformed earthcare from a technical to socio-technical issue. In this context, adaptive management has emerged as a new hope for integrating traditional and scientific knowledge for sustainable natural resource management and earth-care. As a result, increased attention has been paid to improve upon various ways and methods to blend traditional knowledge with frontier technologies and eco-village network. The paper aims to improve our understanding about dynamics, potential and limitations of traditional and scientific knowledge and need to integrate the two through adaptive learning model. First, a short overview of the relevance of traditional knowledge in development is given. Secondly, a conceptual framework in which traditional and scientific knowledge is integrated to achieve sustainable natural resource management. Using data about the forest and water conservation and management, a model of pro-environmental action has been presented as a means to examine integration of knowledge systems for earthcare.

R. L. Shiyani, H. R. Pandya and I. U. Dhruj discuss the methods of quantification and its subsequent economic benefits for conserving an ecosystem. Their case study of Gir forest in Gujarat is empirical. Gujarat has been endowed with plentitude of bio-diversity and natural eco-systems. The Government of Gujarat has declared an area of 1412.12 sq.km as Protected Area, besides 470.50 sq. km as buffer zone. This study attempts to quantify environmental and economic benefits of conserving Gir eco-system. A total of 240 farmers from four different boundaries were surveyed during the year 2000 to examine the impact of Gir forest on agriculture. Increase in income and employment generation through agriculture, hotel industries, tourism, transport, marketing activities, grass collection, sugar industries and collection of timber and firewood as well as conserving bio-diversity and benefits from carbon sequestration are the major advantages of Gir forest. The total economic value of Gir computed by considering various kinds of benefits and also the expenditure was estimated to be Rs. 21105.88 million. This implies that the maintenance of Gir should be accorded high priority by all concerned. Establishment of agro-processing industries in the area nearby Gir will go a long way in uplifting the farmers and other sections of people by way of generating income and employment.

Dewaram A. Nagdeve in his paper on *Population* and Land use: Concerns for Sustainable Development, has mainly explained the close relationship between the three elements mentioned in the titled of the paper. Though the relationship is complex, population size and growth tend to expand and accelerate these human impacts on natural resources and the environment. This paper is an attempt to study the population change, increasing urbanization and its influence on natural resources, and the environment. The analysis reveals that rapid population growth plays an important role in depleting natural resources from deforestation to land degradation, air and water pollution and environmental problems of the country. The considerable magnitude of air and water pollution increases the number of people suffering from respiratory and water borne diseases and many a times leading to deaths and serious health hazards. The analysis reveals that there is an urgent need to control population by helping couples to prevent unintended pregnancies by providing family welfare (planning) services. The importance of population control and environment, and protection of natural resources for sustainable development has been highlighted. The paper concludes with some population policy reflections and emphasizes the potential importance of natural resources for sustainable development. Population and development policy aimed at overall development should certainly include efforts to control population and environmental pollution for sustainable development and for better health of present and future generation.

Narottam Sahoo and Vinu T. J. in their paper on *Policy for Creating Science Literacy Development: A Case of Gujarat*, highlight the significance of better understanding of science and technology and its role in facilitating sustainable development. We face a changing world in which Science and Technology have direct or indirect impacts on the daily lives of people from the most remote village to the major cities. Ignorance of science and technology, formerly considered hazardous and foolish, will in the future become suicidal. Scientific literacy is becoming important for the understanding of the issues of the society and also the expression of democratic rights in this direction. The formal education system cannot alone cope with rapidly changing science scenario. This paper explores on the importance of science literacy for achieving the goal of sustainable development. It also discusses the features of science education and outreach programmes of Gujarat Science City and its methodology for achieving the goal.

Section II: Man and Environment: Sustainability Issues

Man occupies driver's seat in the debate on sustainability. It is an accepted fact that we have reached beyond the possiblistic mode, only to experience the impact of its extremity. In our attempt towards the sustainability, state, civil society, market, science and technology and local institutions, all have important roles to play. During the latter half of the last century, the World economy expanded seven-fold along with disproportionate consumption of material resources and adoption of lavish life styles by the developed countries. International trade also grew rapidly. The economic indicators showed remarkable progress. These developments were, no doubt, different from country to country, and especially between the developed countries and developing countries. But while the economic growth indicators showed progress, the ecological indicators tell a different story, by every ecological yardstick. This is reflected in the current state of our eco-systems that supply our food, i.e. forests, rangelands, fisheries and cropland, as well as the minerals and raw materials that support our industrial economy, marked by mismanagement of natural resources, their reckless exploitation and reduction in our natural resource stocks. Four papers cover the man-nature interface, through case studies.

Narinder Kaur and Sudha Srivastava's paper titled *Dynamics of Man-Environment Relationship in a Tribal Area: A Case Study of Maharashtra*, discusses the problems of scheduled tribes in exclusive and fragile areas due to blanket development policies that do not take into account their traditional knowledge and life style. The tribal communities have been living in the hills and forests in varying concentrations. They can be differentiated from non-tribes on the basis of certain characteristics of habitat, traits like language, beliefs, rituals, customs and institutions. They are very close to nature and for their needs depend upon their immediate environment, sharing an extremely harmonious relationship with it. Though external influences are reaching tribal areas, the level of technology in tribal communities is still low. This paper suggests different development strategies that should be adopted according to the specific needs of the people and their environment.

Surinder Suneja, in his paper on *Ecological Resources and Its Impact on Urban Areas*, has detailed out the methods and techniques to assess and preserve forest and wildlife. This paper, while highlighting the value of forests and wildlife, examines the strategic assessment of conservation and elaborates on the regional environment management plan, habitat and resource inventories and conservation management plan.

S.C. Tiwari and S.S. Singh in their paper on *Influence of Soil Degradation on Soil Properties in Hilly Tropical Forest Soils of Arunachal Pradesh*, have made an attempt to evaluate the influence of degree of soil degradation on microbial biomass carbon, dehydrogenase activity and other soil physico-chemical properties in hilly forest soils experiencing varying degree of forest degradation. This study demonstrates that soil degradation due to jhum cultivation and selective logging caused greater reduction in physico-chemical soil properties, namely bulk density, microbial biomass carbon levels and dehydrogenase activity.

N.C. Jana's paper on *The Role of Geomorphic Impact Assessment in Land Utilisation Studies* explains how the geomorphology of any given landscape determines the land use practices. Landscape, a combined effect of structure, process and stage greatly determines the degree of land use practices. The physical properties such as geology, nature of relief and underground water, drainage conditions, climatic phenomena, soil character, and vegetal cover etc. and their impact of land utilization are dealt with in applied geomorphology. In the present juncture of development-environment nexus the application of geomorphological knowledge in resource appraisal especially the land resources is more relevant to meet the increasing needs and aspirations of the human society.

Section III: Sustainable Livelihoods and Agriculture

Agriculture has strong social and economic dimensions. It is linked to food and nutrition security, poverty alleviation, livelihood opportunities, development of local level institutions and credit institutions. In the rural areas of developing countries, sustainable agriculture is linked with farmers' life style. In the 21st century, sustainable agriculture and livelihoods will require multi-pronged strategies and methods, in order to maintain productivity, equity and equality to the billions around the world. Development of location-specific model watersheds in various agro-ecological zones, precision farming, integrated livestock and farming systems, increasing recycling ratios at the lowest level, efficient linkages between agro and rural based industries and convergence of policies and programmes, are capable of ushering in sustainable livelihoods for the rural poor. Papers in this section touch upon some these issues, through case studies and reviews.

Anil P. Tambay and I. P. Bhagwat's paper *The Jana Utthan Approach: BAIF's Approach to Poverty Alleviation in Rural Areas,* addresses resource utilization for poverty alleviation through micro-level participatory planning approaches. Despite efforts made over the past few decades, rural poverty in India continues to be significant. While the Government and other agencies have strengthened anti-poverty programmes, the numbers of the rural poor have remained static. This paper contains BAIF's approach of poverty alleviation, evolved during their venture with the European Union. This approach has been termed as "Jana Utthan" (People's Upliftment), evolved as part of the Project. It is a "learning approach" to development, which is flexible, can be used to target the very poorest and yet goes beyond being only a "livelihoods approach". The origin of this approach lies in the realization based upon practical experience that the needs of the poorest rural families are diverse and cannot easily be addressed through a standardized model or package of interventions to be applied to all families.

Niladri Ranjan Dash has discussed the issues pertaining to tribals, their habitat and access to sustainable livelihood, in his paper on *Milieu Specific Approach to Sustainable Livelihood.* Spatial and social imbalance in the levels of development poses a great challenge to the country, where a few regions and segments of the population lag much behind others. Since millennia, through age-old social and economic traditions, the tribes had sustained themselves maintaining a delicate balance with their ecologically fragile milieu. With progress of time, penetration of various non-tribal formations into the tribal territories destabilized the tribal social and economic modes and caused erosion of their traditional economic resource base. This paper highlights the drudgery of tribles and advocates to devise measures to individual and distinctively intelligible tribal milieu of the country.

M. Asokhan and G. Ranganathan's paper on *Sustainable Agriculture: Its Goals, Strategies and Concepts—An Analysis* has discussed farming practices and the changing philosophy of agricultural practices. Sustainable agriculture is not merely a collection of discrete farming practices, but is a vision of the future. Its success ultimately depends upon our willingness to accept and implement a new philosophy of agricultural production. Scientists and policy makers around the world are searching for economically sound and environmentally sensitive methods to produce the food, needed for an exploding global population. As we explore socially acceptable ways to bring our human numbers into more realistic balance with physical space and resource limitations, the international community in agriculture is testing alternative farming practices as well as implementing national policies that are designed to assure enough food. This paper also attempts to differentiate between "sustainable," "alternative," "low input," "ecological," and "regenerative" agriculture.

C. R. *Pathak's* paper titled "*Policy Issues for Sustainable Agriculture in West Bengal*", briefly discusses the agricultural development in West Bengal and explores the contextual issues towards a policy framework. Agricultural growth in West Bengal has been brought about not so much by increase in land area but by increasing the cropping intensity and yield per hectare. The high productivity in agriculture has been contributed by the technological inputs, not by expansion of land area. Since small and medium size farmers constitute the majority, the paper dwells into the challenges that these farmers will have to face in the wake of technological changes and structural changes. An attempt has been made here to evolve a suitable policy framework in the light of challenges offered by the uneconomic farm size, technology and input-driven agriculture and the current policy of the state government.

Christy Fernandez, in his paper on *Agro and Rural Industries: Development Strategies*, has examined rural industries in the wake of liberalization, privatization and globalization. Rural industries are backbone for rural employment and income generation and livelihood security. The WTO Agreements have a direct bearing on the agro and rural industries. Competition in the domestic market from goods imported at lower tariffs will be more severe. In the export market, they will face competition from other developing countries and non-trade barriers like environmental standards, occupational and health safety, working condition, labour standards, labour welfare measures, technical barriers like sanitary and phytosanitary measures etc. Appropriate capacity building and skill development measures are urgently warranted along with necessary policy interventions to facilitate them to handle these issues successfully. The specialized organisations both at the centre and the states have a significant role to play in building up capacities of the stakeholders. This paper explores the key for maintaining the competitive edge for these industries, and suggests both policy level interventions and grassroots initiatives.

Satyawati Sharma, Suman Kashyap and Padma Vasudevan, in their paper titled "*Maximisation of Interactions Among Micro and Macroflora for Sustaining Agroforestry in Alkaline Soils*", looks into the scientific explanations of agroforestry in âlkaline soils. Agroforestry is a sustainable land management system combining agri or horticultural crops with tree species with maximum desirable root microbial interactions with the principle of sustained biomass yield and soil conservation and would have the great potential to optimize production and economic return per unit area. Alkaline soils are the sick soils suffering from sodicity/alkalinity (excessive exchangeable sodium content) and are generally poor in fertility with low available nitrogen, phosphorus, potassium, manganese, calcium and copper nutrients. The low fertility, high osmotic pressure, reduced availability of nutrients, poor air and water movement and low microbial activity are the important problems which have a direct bearing on the nutrition of plants on salt affected soils. With anti-plant growth properties, these soils have many adverse effects, e.g. causing low yields of crops or crop failure in extreme cases, limiting the choice of crops, rendering the quality of plants poor, causing excessive run off and flood owing to poor infiltration resulting damage to crops in the adjoining area.

Section IV: Water Resources Management

Fresh water is becoming one of the scarcest resources of the earth on which the entire life support system thrive. Management of water resources depends on ground water development, efficient utilization for irrigation and its provision as a social good. Ground water is the most ill-managed among various sources of water. The adverse effects of ground water mismanagement are serious in terms of equity and sustainability of the resource due to the negative externalities associated. Water resource conservation is an important component of water management. Efficient irrigation management, monitoring of ground water's quality and quantity and rainwater harvesting are the driving forces behind water resources management. This section throws light on some of these issues.

C.B.S.R. Sharma's paper *Restoration of Tanks for Water and Ecological Security: An Ecosystem Approach*, discusses one of the traditional methods for water harvesting that has positive ecological impact. Conservation of water by every conceivable approach has emerged as a great social concern because of increasing scarcities despite the unaltered quantum of rainfall. Ecological knowledge and wisdoms suggest the possibility of some new approaches. Hindsight wisdom and our pioneering studies lead us to treat the tank as an ecosystem and all things around as its constituents—tank bed and shorelines, banks and slopes, sluices, inlet and outlet channels, human settlements, flora, fauna, husbanded animals, soil and water. The above approach enhances the resourcefulness of the habitats and biota, opening up newer avenues for development by the initiatives of the local people for their own security and welfare. Tanks suddenly emerge multipurpose. The role of Panchayats becomes critical in the management of this system.

R.S. Goel, V.B. Patel and G.P. Mathur briefly highlight the environmental impacts of irrigation management. The environmental impacts of floods, droughts and water storage projects are also briefly summarized together with the role of irrigation in Indian Agriculture. India has 15 per cent of the world's population, but only 4 per cent of the world's water supply with 2 per cent land area. Integrated water management is most vital for poverty reduction, environmental sustenance and sustainable economic development in India. Availability of water in India is under tremendous stress to meet the requirement of food, fodder and fibre for the bulging population, rapid urbanization, increase in per capita consumption, industrial growth and demands for maintaining ecology. Due to the large temporal variations in river flows, storage of water becomes inevitable in Indian monsoonic climate.

Shailza Verma, Renu Bhargava and P.K. Garg focus on drinking water supply problem in rural areas, and suggest the measures through sources selection and utilization to minimizing the problem of drinking water. The work has been carried out in GIS environments for easy handling and analysis of the data needed for the water supply. The exploitation of natural resources is bound to increase to meet the demands of growing population. The exploitation focusing only on economics, without considering ecological consequences, has placed existing natural resources in a critical condition. In order to ensure the availability of sufficient quantity of good quality water, it becomes essential to plan and build sustainable water supply schemes, which may provide potable water according to demands.

Contents

IV
WATER RESOURCES MANAGEMENT

List of Tables

List of Figures

List of the Contributors

Anil P. Tambay, BAIF Development Research Foundation, Dr. Manibhai Desai Nagar, Warje, Pune-411052.
E-mail: baif@vsnl.com

C.B.S.R. Sharma, Salim Ali School of Ecology and Environmental Sciences, Pondicherry University, Pondicherry-605014.

C.R. Pathak, Secretary, Regional Science Association, India, CK 134, Salt Lake City, Sector II Calcutta-700091, West Bengal.
Email; cr pathak@hotmail.com

Christy Fernandez, Chairman, Coir Board, Department of Agro and Rural Industries, Government of India.

Dewaram A. Nagdeve, Senior Lecturer, International Institute for Population, Sciences, Govandi Station Road, Deonar, Mumbai-88. E-mail: dnagdeve@yahoo.com

G. Ranganathan, Professor and Head, Department of Social Sciences and Languages, Anbil Dharmalingam Agricultural College, and Research Institute, Trichy-620009.

G.P. Mathur, Deputy Director, Environmental Management, Central Water Commission, 413(S) Sewa Bhavan, R.K. Puram, New Delhi-110066.

H.R. Pandya, Department of Agricultural Economics, Gujarat Agricultural University, Junagadh-362001.
E-Mail : rlshiyani@,gaui.gui.nic.in

I.P. Bhagwat, BAIF Development Research Foundation, Dr. Manibhai Desai Nagar, Warje, Pune-411052.

I.U. Dhruj, Department of Agricultural Economics, Gujarat Agricultural University, Junagadh-362001.
E-Mail : rishiyani@gaui.guj.nic.in

K.V. Sundaram, Chairman Emenitus Bhoovigyan Vikas Foundation, New Delhi-110070.

L. Nirmala, Research Associate (Agricultural, Extension), Anbil Dharmalingam Agricultural College and Research Institute, Trichy-620009.

M. Asokhan, Assistant Professor (Agricultural, Extension), Anbil Dharmalingam Agricultural College and Research Institute, Trichy-620009.

M. Moni, Deputy Director General, National Information Centre Government of India, CGO Complex, New Delhi-3.

Mrityunjay M, Jha, Consultant, SMEC India, Gurgaon-8.
E-mail: mrityunjayjha@yahoo.com

N.C. Jana, Centre for the Study of Regional, Development, School of Social Sciences, Jawaharlal Nehru University, New Delhi - 110067.

Narinder Kaur, Research Student, Department of Geography, University of Mumbai, Mumbai-400098.

Narottam Sahoo, Gujarat Council of Science City, Block 11, 9th Floor, Udyog Bhavan, Gandhinagar-382017, Gujarat, India.
E-mail: narottam@scity.org

Nidhi Sharma, Project Associate, Indian Institute of Forest Management, Bhopal, India.

Niladri Ranjan Dash, Department of Geography, Faculty of Science, The Maharaja Sayajirao University of Baroda, Vadodara-390002, Gujarat, India. E-mail: niladridash@eth.net

P.K. Garg, Associate Professor, Civil Engineering Department, Indian Institute of Technology, Roorkee-247667.

Padma Vasudevan, Centre for Rural Development & Technology, Indian Institute of Technology, Hauz Khas, New Delhi-110016 (INDIA).

R. S. Goel, Member Secretary, National Environmental Monitoring Committee, for River Valley Projects. 413(S) Sewa Bhavan, R.K. Puram, New Delhi-110066. E-mail: goelrscwcyahoo.com

R.L. Shiyani, Department of Agricultural Economics, Gujarat Agricultural University, Junagadh-362001.
E-Mail : rlshivani@gaui.gui.nic.in

Rekha Singhal, Associate Professor, Indian Institute of Forest Management, Bhopal, India, *E-mail*: rekha@iifm.org

Renu Bhargava, Professor, Civil Engineering Department, Indian Institute of Technology, Roorkee-247667.

S. C. Tiwari, Department of Forestry, North Eastern Regional Institute of Science and Technology, Nirjuli-791109 (Itanagar), Arunachal Pradesh, India.

S. S. Singh, Department of Forestry, North Eastern Regional Institute of Science and Technology, Nirjuli-791109 (Itanagar), Arunachal Pradesh, India.

Satyawati Sharma, Centre for Rural Development & Technology, Indian Institute of Technology, Hauz Khas, New Delhi-1110016 (INDIA),
E-mail-satyawatis@hotmail.com

Shailza Verma, Research Assistant, Civil Engineering Department, Indian Institute of Technology, Roorkee-247667.

Sudha Srivastava, Reader, Department of Geography, University of Mumbai, Mumbai-400098.

Suman Kashyap, Centre for Rural Development & Technology, India Institute of Technology Hauz Khas, New Delhi-1110016 (INDIA).

Surinder Suneja, Professor & Head, School of Planning & Architecture, I. P. Estate, New Delhi-110 002.

T.J. Vinu, Gujarat Council of Science City, Block 11, 9th Floor, Udyog Bhavan, Gandhinagar-382017, Gujarat, India.

V.B. Patel, Chairman, Board of Governors, Water Management Forum, The Institution of Engineers (India), 128, Manekbaug Society, Ambawadi, Ahmedabad-380052.
E-mail- vbpatel@multimantech.com

I
Earthcare Policy

1

The Case for an Earthcare Policy Institute in India

K. V. SUNDARAM and M. MONI

Introduction

The Common Minimum Programme of the UPA Government has emphasized the twin objectives of economic growth and environmental conservation, with special programmes for dryland farming, watershed and wasteland development, water management in all its aspects and continued growth of agriculture, horticulture, aquaculture, afforestation, dairying and agro-processing, along with a comprehensive medium-term strategy for food and nutrition security. In a broad perspective, it aims at a truly sustainable development process and implies the need for taking integrated and coordinated efforts to protect as well as achieve the rational use and management of our natural and environmental resources to halt the on-going destruction of Earth's capacity to support life, while at the same time putting the Indian economy at a higher pedestal.

The new century is the "*Earthcare and Knowledge Century*", and many national and global issues will come to a head during this period, calling for a reappraisal of our development priorities, as well as taking some bold initiatives in the policy and planning realms. All these would call for a deeply-held commitment to a new ethic, *the ethic for Earthcare, sustainable development, sustainable consumption* and *life styles*. How to translate these principles into practice by working through the systems already in place in the country, is our major dilemma. This paper is an attempt to outline a proposal to fulfil the development policy requirements of our government, within the framework of a strategy for caring for the Earth and achieving sustainable development. Its acceptance and implementation would go a long way in ushering in a new era of development without destruction.

Towards Earthcare Policy

Earthcare implies a development paradigm, wherein issues in environment and economics are circumscribed by the requirements of human development and ecological sustainability. It is an overarching concept. It includes besides conservation of earth's vitality and diversity, human actions to minimize the depletion of non-renewable resources, keeping development actions within the carrying capacity of the earth, providing a rational framework for integrating development and conservation, changing personal attitudes and practices to enable communities to care for their own environments and ensuring respect for the preservation and care of human culture and improving the quality of human life. The multitude of objectives inherent in this statement makes it amply clear that it is a multi-dimensional and multi-disciplinary concept, requiring an integrated and coordinated approach by governments, in which effective institutional mechanisms will have a crucial role. While there are institutions to take care of issues pertaining to human development, there are none to take a holistic and integrated view on the conservation and strengthening of earth's vitality, on which all lives on planet depend. As most of the important issues to be addressed in this context are strongly interlinked, our actions must be mutually supportive, synergistic and aimed at the common goal.

The Present Imperatives

The state of natural resources in our country and the extent of degradation and depletion due to human interference are alarming, to say the least. Without going into the statistical dimensions, we may list a range of issues to be addressed to reverse the deterioration of the earth. They will include problems like: *land degradation*;

deforestation; water scarcity; quality deterioration and pollution of water; air pollution; loss of non-renewable energy resources; over-pumping of aquifers leading to falling water tables, loss of biological diversity, degradation of the coastal ecosystems and wetlands etc.

Natural resource use and management in the country is at present a highly compartmentalized activity, with different departments, institutions and other stakeholders, operating in their restricted jurisdictions. The matter is further compounded by trade-offs among the conflicting goals of the biological, economic and social systems. This has often led to confusion, greater misuse of natural resources and overall resources depletion. As long as such confusions exist, sustainable development of natural resources will remain an elusive goal. There is an urgent need *to look at Earth Systems* (Comprising the lithosphere, hydrosphere, biosphere and the atmosphere) *in their totality* and comprehensiveness, and view the problems in an integrated and holistic manner. Only an "*umbrella organization*", at the National level, in the nature of a cross-sectoral mechanism, can handle this difficult task. Such an organization can make a substantial contribution to integrating sound natural resource management principles into all levels of policy-making in the country, leading to the establishment of appropriate institutions and extension systems at the regional and local levels. We need not labour more on the point that caring for the Earth and achieving sustainable development through a strategy of mutually reinforcing actions at individual, local, regional and national levels is *a highly balancing act*, and calls for creating a multi-layered integration of management systems, for which appropriate institutional arrangements are necessary. The compulsions of the new "*Earthcare and Knowledge Century*" and the present political scenario in the country demands such a comprehensive approach to ensure unified action to preserve the earth systems to achieve sustainable development, consumption and life styles. In this context, a National Policy Framework for resource-use and management would also be an essential instrument.

A Stock-Taking of Our Institutional Structures

Against the above background, let us now take a close look at our current national framework for integrating development and conservation. We have already a fair administrative and technical set up at the national and sub-national levels, embracing economic policies, national laws and regulations and a fairly good information base, which however needs to be improved in the light of the new developments in Information and Communication Technology. The need for Decision Support Systems at the micro-level has now taken on a new meaning with the current overwhelming emphasis on micro-level planning. Sectoral data on natural resource endowments in the fields of Geology, Geomorphology, Land Use/Land Cover, Water Resources, Soil Resources, Waste land, Fisheries, Forestry etc., at the *Village Level Resolution* are now essential, for establishing a sound Development Planning and Decision-making system for sustainable development, extending from the national to the grassroots level.

Following the *Brundtland Commission Report*, the National Government has also established a Ministry of Environment at the Centre, along with similar action at sub-national levels. These institutional structures are sectoral bodies with limited mandates. They primarily deal with some environmental policies, which fall very much short of the more comprehensive and holistic Earthcare Policies that seek to keep our spaceship-earth-inhabitable. Their working has been generally reactive and programme-driven, responding to problems after they have surfaced and when they are more expensive to treat, than if they had been tackled much earlier. Their links with the numerous resource management agencies are often weak. The Planning Commission, *without an independent policy analysis and research group* to formulate and back- up its plans and policies, has been mechanically assembling the policies emerging from these fragmented ministries and organizations. It is *the sectoralism in this planning process*, which has obscured a broader vision for shaping the wider policies concerning the earth.

The Organisational Vacuum

The basic dilemma at this juncture are two-fold:

(i) Who will *integrate and coordinate* the splintered and fragmented policies, emerging from the Government Ministries and Departments?
(ii) Who will *screen, filter and distill* the research findings, emerging from the few earth sciences departments in universities and

some independent research institutions, which are conducting limited fundamental and applied research, relating to earthcare?

The answer to the above questions is negative. As for integration of policy functions, the Planning Commission had earlier in the 1960s, a separate Division for natural resources, which was later abolished, when individual natural resources got separated in various independent ministries. *At present, there is no policy research organisation in the country which can address itself to these issues*. As a result, fragmented policies and plan programmes have continued to emerge, which do not relate to each other, sometimes overlap among themselves and perhaps even negate each other.

Issues in Filling the "Organisational Vacuum"

In this paper, we are addressing the question of how to re-organise and strengthen the national institutional policy-making mechanism in the country, particularly in relation to Earth systems planning and sustainable development of natural resources, which are the crucial challenges for the first decade of this new century.

Any solution that we devise for filling the *organisational vacuum* indicated above, must be by creating a mediating and *empowered quality organisation*, which can actively interface with the already existing pluralistic and sectorally-oriented policy-making structures in government and pave the way for the coordination and integration of policies in the Earth Sciences sector. It should be designed in a manner that it uses the best human resources, wherever available, through internal and external networks, partnerships and alliances.

Successful policy formulation in any sector of activity would rest on a shared knowledge from several "*feeder institutions*". This implies systematic flow *of* (a) *information* from data-base systems, (b) *of knowledge* from fundamental and applied research institutions and (c) *of people's perceptions* from area-specific studies and field observations at various levels. The research findings from various organizations, having different policy implications, need to be collated, verified, filtered and distilled into appropriate policy formulations. The proposed institution will keep abreast of the long-term policies, to position itself as a major contributor to the identification of issues and priorities as well as solutions. Its success will largely depend on close networking, cooperation, teamwork and shared knowledge with not only government organisations, but also with the various institutions performing fundamental and applied research in the country.

In the area of Earth Sciences Research, we have a few Central Government Institutions, undertaking research related to natural resources and their management, like the ICAR, CSIR, Department of Science and Technology, Institute of Ocean Development, Centre for Meteorological Studies, Department of Space, etc. We have also an institution called the Centre for Earth Science Studies (CESS), set-up by the State Government of Kerala in Trivandrum, which is an earth science related research oriented organization. A comparative analysis of the functions and focus of some of the relevant institutions makes it amply clear that none of these really cover policy research issues concerning the holistic and systemic view of earthcare.

The above-mentioned institutions are either uni-disciplinary or problem-oriented research bodies. They do not specifically address themselves to formulation of integrated Earthcare Policies, which are National as well as global in scope. Among the new organizations, which have cropped-up in recent years, the *Bhoovigyan Vikas Foundation (BVF)* is one, which is aiming to be a policy-oriented research organization in the Earth Sciences sector. The BVF has been conceived as a Consortium of Earth Science Professionals and is just five years old. But it is fast moving towards, both in its mission, objectives, programmes and activities, into a policy-oriented research organization, dealing with the wider issues of National and Global importance, related to Earthcare and Sustainable Development. By suitably enlarging its policy-research dimensions and building up its infrastructure, it can fill the organizational vacuum.

During the 5 years since its inception, it has established its credentials across the country, for its commitment to earthcare movement in India. It was launched with the celebration of the First Earth Day of this millennium and is a Member of the World Earth Day Network. It has successfully conducted two International Conferences, the first one on *Sustainable Development and Sustainable Life Styles* and the second one on *Sustainable Agriculture, Water Resources Development and Earth Care Policies*, besides one colloquim and one training programme. It has brought out 5 publications, with three more in pipeline. Further, it is engaged in research projects for the Department of Science and Technology and the Council for the Advancement of People's Action and Rural Technology (CAPART). Some of its research proposals with national and international agencies are in pipeline. Among other activities, the BVF's advocacy for the establishment of 'Open Air Museums with Earth Care Message', has led to the establishment of an open air museum of "Earth, Man and Environment" in Raipur (Chhattisgarh) over an extensive campus of 350 acres of land, which is being implemented. It has also assisted in preparing a Master Plan for this Museum.

The Foundation created a niche for itself among the scientific community by honouring luminaries in the Earth-related sciences like Dr. K. Kasturirangan, Dr. N. R. Narayana Murthy, Prof. R. P. Misra, Prof. Mohammad Shafi, Dr. N. C. Gautam and Prof. G. S. Bhalla for their outstanding scientific contributions. During a short span of time, it has established its visibility and image at the national and international levels. It has 3 local chapters and one Overseas Chapter in Canada. It also won accreditation for participation in the Johannesburg Meet (World Summit on Sustainable Development) in 2002. It has also established collaborations with the United Nations Centre for Regional Development, Nagoya, Japan; Asian Institute of Technology, Bangkok, Thailand; and the Earth Policy Institute in Washington, U.S.A. *With a minimal start-up grant and a recurring annual working grant, it can easily blossom into a leading Earth Policy Research Institute of National and International Importance.*

The Proposal for an Earthcare Policy Institute

It is in this context that we propose an *Earthcare Policy Institute (EPI)* to be established as a National Centre of Excellence. As a policy research and analysis centre, the central activity of EPI will not only be research and policy formulation (RPF), but also other related scientific activities (RSA), such as scientific data collection, information services, policy feasibility analysis, and education support (training). Through its programmes, activities and other professional relationships with partners, including government, academia, industries and other stakeholders, this centre will function as a *key institution*, contributing to the formulation of innovative policies in the Earth Sciences sector.

Mission

The proposed *EPI*, is intended to promote an integrated and holistic view of Earth Systems management, as well as the conservation and development of natural resources and *advocate appropriate policies* for Earthcare and sustainable development of natural resources on a continuing basis. It is proposed to be developed as a *Centre of Excellence*, devoted to creating initiatives for policy-oriented and advanced multi-disciplinary research studies, bordering on the realms of various Earth Sciences and their frontier areas. With a deeply-held commitment to the ethic of earthcare, sustainable development and sustainable living, it will seek to translate its principles into practice. It would also seek to network, and interface with the world of academia and the world of public policy, by serving as a bridging "*Think-tank*", with an intensive advocacy role and interfacing with all policy-moulding institutions in the country, as well as in the developing and developed countries abroad. It will provide a forum for creative dialogue among policy-makers and representatives of earth-related research and economic and social development research organizations, as well as leading NGOs. It will act as the storing and clearing-house for the exchange and constant flow of information, ideas and activities, relevant to earthcare, and complementing the efforts of several government and non-government bodies in the country, with its shared perceptions. From the research and scientific points of view, its dimensions of activities and linkages will be both national and international.

Role, Functions and Activity Components

The *EPI* is envisaged as an institution, functioning with flexibility and openness. Its broad areas of focus will comprise:

(i) Advocacy and Awareness Building,
(ii) Policy Analysis and Research, including Action Research,
(iii) Training,
(iv) Advisory/Consultative services, and
(v) Networking and information exchange.

The *EPI* would also undertake some relevant fundamental and field-oriented Action Research, as well as process the numerous research contributions generated at other specialized research institutions. Visualized as a 'think-tank' and a 'change agent', the *EPI* has to build on the *research inputs* from the technical and professional academic institutions, working in the areas of emerging science and technology, and also on *grassroots level experiences* of Non-Governmental organizations. This implies active partnership and collaboration with all those concerned with our '*common future*', in which each will gain from the interaction and will also contribute something to that interaction. Thus the contributions emerging from the *EPI* will be a combination of practice with scholarship, experience with vision, and pragmatism with policy directives. In all such activities, the Centre will play a *catalytic role* in bringing together, a wide range of actors linked by their common interests in earthcare, sustainable development of natural resources and related research issues.

The *EPI*, as a policy-oriented research institution will adopt a '*Systems Engineering Approach*' to the solution of environmental and earthcare problems requiring an inter-disciplinary and multi-lateral approach involving both physical and social scientists and including studies on rural, urban, natural and human resource development problems. For this purpose, the *National Centre* will draw upon the intellectual resources available in Earth and Social Sciences Departments of Universities and Research Institutes, specializing in Earth-related and Planning-oriented studies. For this purpose, the *EPI* will have three major groups or divisions with their distinct responsibilities (See Figure 1.1):

(i) One Division will undertake selected *Thematic and Policy-oriented Studies* with four departments:

   (a) Natural Resources Management, Sustainable Development/Consumption/Life Styles.*
   (b) Geomatics applications and future studies.
   (c) Surveys and Applied Research.
   (d) Studies on Earth and Human Systems and Anthropogenic interventions.

---

* Studies on climate change, ocean development and energy, are currently being undertaken by some specialized institutions. EPI will not enter into their arena, but will draw upon their research findings for policy formulation.

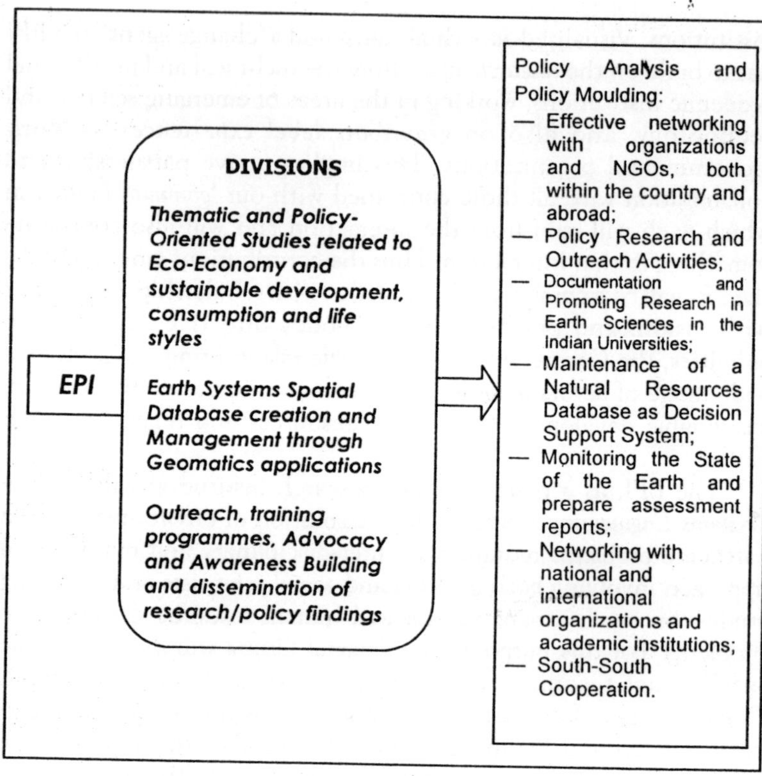

Figure 1.1: Proposed Organisational Structure of EPI

(ii) A *Spatial Database and Management Division* of the Centre, which will be continuously documenting all relevant data and information about the status of health and well-being of the Earth in its various eco-systems and will be assisting and facilitating all the research activities of the Centre.

(iii) The third major division of *EPI* will be an *Outreach Division*, which will actively establish links with all academic and research institutions in the country engaged in "*Earth Science Studies and Research*". This Division will effectively network with these research institutions, screen, filter and distil their research findings and use them in planning and policy formulation. The *Outreach Division*, while performing its coordinating function will also perceive and identify the areas, where study and research activities are lagging, and seek to promote such research activities, either in one of the research divisions of *EPI* or by outsourcing such research to other institutions, which have the desired capability. For this purpose, the *Outreach Division* in *EPI* will operate a *Research Promotion and Coordination Budget* and will supervise this activity. This division will also establish links with the surrounding countries of the region and promote a *South-South Cooperation*. Conducting training programmes for human resource development is another branch of its activities.

Thus the major activity components of *EPI* will include the following:

(i) Networking with not only governments, academia, and relevant NGOs in the country, but also with international institutions abroad and initiating South-South Cooperation.
(ii) Advocacy and Awareness Building through print media and celebrations like the Earth Day and Promoting Earthcare by People's Groups;
(iii) Conducting Participatory Action Research to enhance the understanding of various critical and emerging issues through case studies and development cooperation/collaboration with governments and NGO's working in specialized areas;
(iv) Creating spatial data base systems (GIS) in an electronic network for information exchange and dissemination of research results;
(v) Training on various subjects related to natural resources development, management and conservation, land use, sustainable development of rural and urban areas, human development etc.

In this context, networking, coordination and constant interaction will constitute the major planks of its activities. It will actively liaison with the United Nations and its agencies, interface with the Ministries of the Government of India and all State Governments, and collaborate and network with the leading International and National NGOs on the relevant issues. The networking functions in detail may be seen in Figure 1.2.

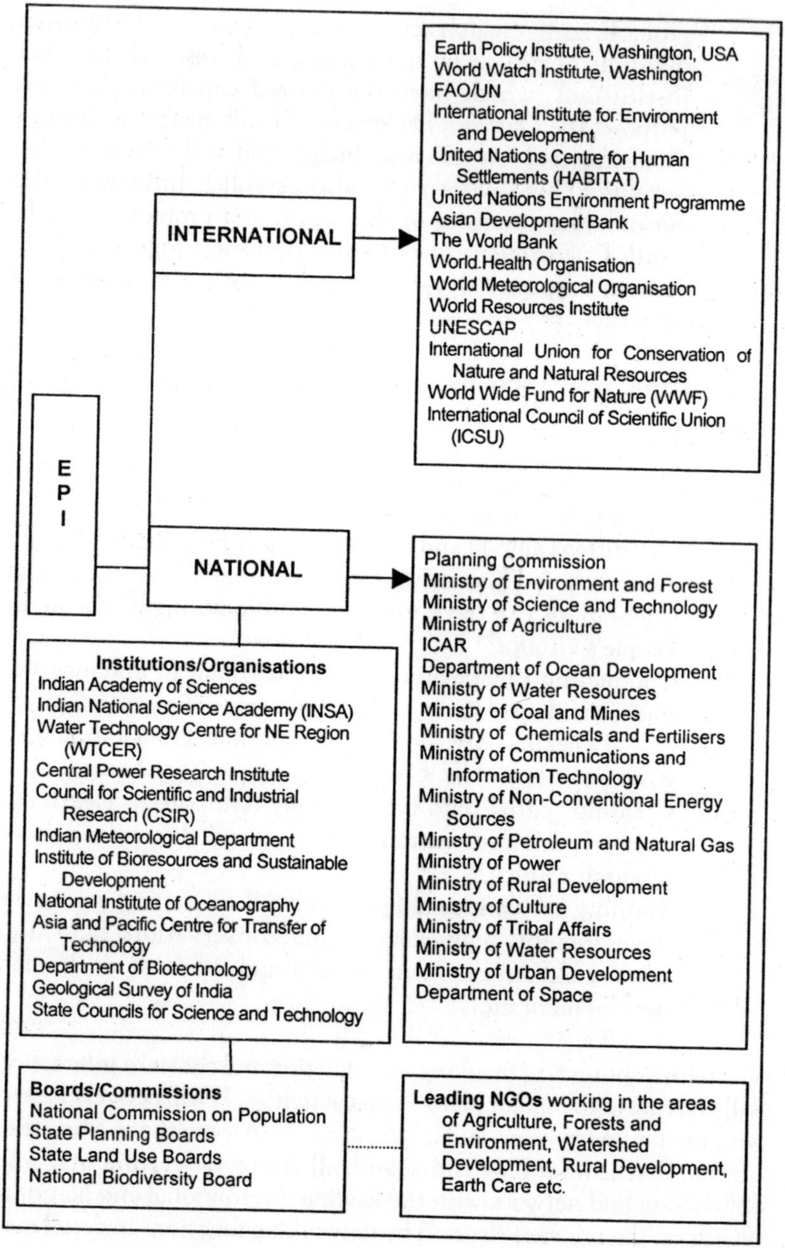

Figure 1.2: Networking Functions of EPI

With its activity components, as proposed above, the *EPI* will have the potential to emerge as a "*High Performing Knowledge Institution*", establishing in due course, linkages with surrounding countries and emerging as the *Regional Centre for South-South Cooperation*. Thus it will help evolve a new era of *Ecological Diplomacy*.

2

Population and Environment: Facing the Challenges

K. V. SUNDARAM and MRITYUNJAY M. JHA

Introduction

Throughout the developing world, population is growing, while per capita arable land and water resources are shrinking. In spite of vast developments in *Technology and Development* techniques, the developing countries of the world today are in a deep-seated and multi-faceted development crisis, mainly characterized by *three major challenges*: i.e. (i) *Population explosion,* (ii) *Environmental degradation*, and (iii) *Economic stagnation*, which is a consequence of the above two.

The Population Explosion

Among the various factors influencing the development process, the potential population explosion is almost a nightmare. It has tremendous repercussions on every Nation's physical resources and social services. Even as population continues to grow, the habitable portion of the planet is shrinking, owing to ecological degradation, and the rural population is bound to face greater and greater land scarcity. Whole families will be constrained to subsist on a mere one hectare of land or even less. Poverty will reach unimaginable dimensions. This is the scenario that emerges from a global analysis of the historical trend of population growth in the developing countries, as visualized by many World Organizations.

The well-known African Leader, Robert S. McNamara, in an address to the *Africa Leadership Forum*, made a very distressing statement, which I quote: "In these countries, population is growing exponentially, with a sharp decline of the extent of arable land per person to less than 0.3 hectares, with acute food insecurity and hunger, resulting in famine conditions, nutritional deficiency and hunger and leading to migration and even death. A child born today in Sub-Saharan Africa faces a bleak set of statistics. The odds are 1 in 10, that he or she will not live for more than a year and the child's mother will die giving birth. As for life expectancy, the child can expect to live for only 50 years –25 years less than children in the developed countries" (McNamara, 1984).

In India is not so bad; but still we face the problems of food insecurity, nutritional deficiency and hunger, in a context of regional inequalities, sharp social differentiations and problems relating to the degradation of our natural resource base. Almost every year, this vicious cycle of under-development spins in different ways and at different speeds, in different regions of our subcontinent. Despite some achievements in population control, the nexus between poverty and environmental degradation poses a critical threat to our development in this twenty first century.

The Environment and the Ecological Yardsticks

From population problems, let me come to the environmental issues and challenges facing us in this century. The quality of the environment depends on man's keen perception and careful utilization of natural resources. When technology remained primitive and population was comparatively small in relation to land area, pressures on the environment, in terms of its capacity to meet the competing resource needs were not great. But conditions have changed.

During the latter half of the last century, the World economy expanded seven-fold along with disproportionate consumption of material resources and adoption of lavish life styles by the developed countries. International trade also grew rapidly. The economic indicators showed remarkable progress. These developments were, no doubt, different from country to country, and especially between the developed countries and developing countries. But while the economic growth indicators showed progress, the ecological indicators tell a different story, by *every ecological yardstick.* This is reflected in the current state of our eco-systems that supply our food, i.e. forests, rangelands, fisheries and cropland, as well as the minerals and raw materials that support our industrial economy, marked by mismanagement of natural resources, their reckless exploitation and reduction in our natural resource stocks. Today, at the beginning of this twenty first century, we commence our journey on the planet, with a dubious legacy, i.e. with a heavily depleted natural capital base and a highly deficit *'ecological balance sheet'*.

The "Tragedy of the Commons" and a Belated Realization

Let us take stock of what the human beings have done to the Earth. It is a story of *colossal environmental destruction*, to which we woke up only in the early 1970s. This led to a late realization of the need to work towards the common goal of *Saving our Planet and Saving our Progeny*. Earlier in 1798, Thomas Malthus had warned that population, when unchecked, will outrace the Earth's ability to provide sufficient food (Malthus, 1798). Although the neo-Malthusians have sought to soften the argument by holding out the possibilities of effective population control and technological innovations, with our natural resource base under *constant* threat, the case for the revival of our plundered planet, seems weak, atleast in the short-run.

The Triple Bottom Line for Sustainable Action

As human numbers and the scale of their activities increased, the pressure on fragile eco-systems also increased. At the global level, several International conferences have been discussing these problems, since the first one at Stockholm in 1972. The real breakthrough, both conceptually and on the policy front, came in 1987 when the World Commission on Environment and Development (The Brundtland Commission), called for a new style of development that meets "the needs of the present generation, without compromising the needs of the future generations". More conferences followed, one in 1992 at Rio de Janeiro, in 2001 at New Delhi (BVF), and another in 2002 at Johannesburg. The greatest contribution of all these multilateral discussions was in highlighting the need to simultaneously address developmental and environmental imperatives and in defining developmental objectives along a *'Triple Bottom Line'*, i.e. *Economic, Social and Ecological*, which is now referred to by planners as *Inclusive*

Development. The Triple Bottom Line and the objectives implied therein are indicated diagrammatically as follows:

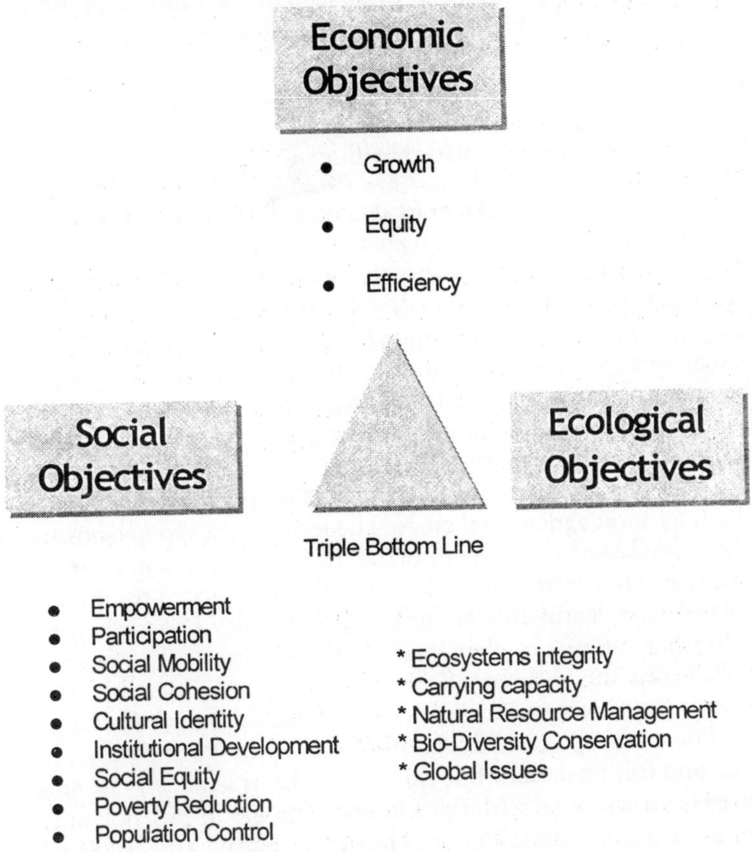

Triple Bottom Line

It became clear that without better environment stewardship, development will be undermined, and without accelerated development in poor countries, environmental policies will fail.

The Role of Technology and an Experience in Transition from Sustainability to Unsustainability

At this stage, physical scientists responded by devising *new technologies* that could ameliorate problems caused by the narrow focus on economic growth and productivity. According to them, technology would solve all these problems. But this was not true. Technology also led to some adverse results, as evidenced by the third generation problems caused by the Green Revolution. This is evidenced by—

- Soils and Streams contaminated by fertilizers.
- Air (Atmospheric Systems) polluted with wastes.
- Crops and forests ravaged by disease etc.

Now the question is what can we do about the problems that result from the intensification of resource management and the expanding scale that often accompanies it? An alternative view in this context suggests that reverting to traditional science and technology, i.e. *Indigenous Knowledge* or *Traditional Knowledge and Practices* will be rewarding. This means that whatever is valuable in this knowledge should be brought into the mainstream concerns and programmes of development, thereby enriching both the modern and the traditional systems.

In the Sahel region of Africa (now Mali and Niger), the climate is very dry with average rainfall less than 600mm per year, which is a lower limit for most agricultural crops. The local ethnic group living in this region called *Marka,* are experts in the cultivation of certain varieties of native rice. They have been cultivating this crop since pre-historic times. They have learnt this art to perfection. They know when to plant and what varieties to plant, taking note of climatic fluctuations. Their decisions are influenced by environmental clues and the performance of different varieties of rice, which have different vegetative periods, different adaptations to various flood depths, flood timing, *p*H tolerance and fish predation. But this knowledge is kept as *secret* and known only to the select Marka Group. This group has not only maintained a specific ethnic identity, but also developed a special *Social Relations system* with other groups and devised some *buffering mechanisms* to tide over the potential bad times. Into these regions, through International aid, a Regional Development Project was introduced, which brought about a change of cultivation from indigenous rice varieties to the Asian variety *Oryza sativa,* which is more marketable and has a higher yield, but requires considerable amounts of water. This proved to be a disaster. As a result of this experimentation, the traditional system got eliminated and along with it the *Secret environmental and social knowledge of the community* (Jordan, 1998). A large portion of the native groups had to abandon the cultivation system and became poorer.

It is only to indicate that technology itself has to adapt to traditional knowledge, where such knowledge has its own intrinsic merits, which can be tapped, protected and improved with innovations and practices.

Implementation Aspects

(a) *The 'Adds-On' Versus 'Integrated' Approach*: A '*holistic, cooperative, multilateral and partnership approach*', needs to be taken to halt the looming ecological and other catastrophies, with due deference to ethics, values and the promotion of justice and equity in consumption and life styles. Persistent poverty, environmental degradation, deforestation and desertification can no longer be ignored. The challenges and trends in this direction can only be reversed by *concerted action*. They call for appropriate planning interventions at the local, regional, national and international levels. A *unified approach to planning and decision-making* is called for by the nations of the world. This implies a vital change or departures from the current operating procedures and programme planning. Many countries have sought to address environmental concerns as "*adds-on*" to development policy-making. This means a *casual approach to meet 'environmental concerns'*, after the core of decision-making has been made. The difference between adds-on and integrated approach is that in the former, 1 + 1 is equal to 2, but in the latter, it is 11.

(b) *Restructure the Planning Mode at Different Spatial Levels*: The adoption of the '*Integrated Approach*' will also correct a fundamental error in our planning methodology in which the costs of environmental and social damage are not embedded in the decision-making process from the beginning. This correction in methodology implies that we have to prepare '*systematic area plans*' at the regional and local levels, besides the macro planning at the national level. The two kinds of planning at the sub-national levels that the Bhoovigyan Vikas Foundation (BVF) has been advocating in this context are:

   (i) Holistic Village Cluster Planning at the local level, with participatory approach as the basis; and
   (ii) Regional Economic-*cum*-Environmental Planning at the regional level, with ecosystems approach as the basis.

For both, the BVF has methodologies, which will help in setting appropriate regional and local priorities, promoting positive synergies to promote adverse environmental impacts and mainstreaming the environmental issues in national policy planning. To achieve this, strong political will and determined action are needed, as these are radical departures in planning. Academic and professional organizations have a great role in devising area-specific methodologies and operationalizing them. They should relentlessly advocate for these drastic changes and reforms in development planning.

We need to deal not only with domestic or national issues, but also with international concerns, like climate change, rising sea level, dying coral reefs and disappearing species. In this respect, humanity, now more than ever, truly stands at a *defining moment* in history. The tide in the affairs of '*Earth and Man*' today beckons us for greater global cooperation and for more cooperative decision-making. *There is not much time left for the planet to be saved.*

Is Development Sustainable?: Some Relevant Postulates

"Where there is no vision, people perish"

This is a well-known planning axiom. The evidence at our disposal clearly shows trends, which mark an increasingly stressed relationship between our national economies and the earth's ecosystems, which are taking a growing economic toll. Sustainable development calls for a radical transformation of the environment and argues that environmental problems need to be considered within the global economic system and that the degradation of the environment is not *natural*, but an historical process linked to economic and political structures. In developing countries like ours, environmental management is no doubt assuming more importance as the contradictions of development—desertification, deforestation, and food insecurity are becoming more and more significant. Providing a vision of an environmentally sustainable economy, along with a road map to reach this goal is undoubtedly a big paradigmic shift involving vital changes in the mindset of the population. Many thinkers have suggested some solutions. But they are all either in the nature of concepts, perspectives or models. I can only cite a few:

(i) Eco-development within the carrying capacity of the Earth with *five major components* (Figure 2.1), which is a valid concept, but whose methodology has to be defined within each national and sub-national framework.

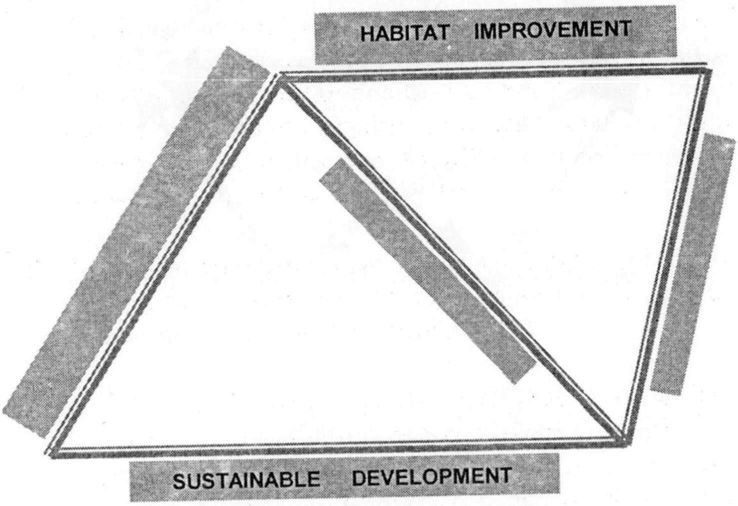

Figure 2.1: Eco Development Model

(ii) A systems approach to planning called "*Co-evolutionary approach*" (Figure 2.2) in which Knowledge, Values, Resources, Technology and Social Organization have been identified as the crucial sub-systems that would support one another. Each sub-system is related to all the others, but each is changing and affecting the evolution of the others. No sub-system will dominate another.
They will all co-evolve to achieve sustainable development. This view has been expressed by Nogaard, who himself considers this as only a '*perspective*' and not really a model.

(iii) A model of planning for an "*Eco-economy*" advocated by Lester Brown (Figure 2.3), President of *The Earth Institute* in Washington. Brown has described how to restructure the global economy to make it compatible with the Earth's eco-system, so that economic progress can continue for ever. In this new eco-economy visualized by him, wind energy will replace coal-mines, hydrogen-powered fuel cells will replace internal consumption engines and cities will be designed for people, and not for cars.

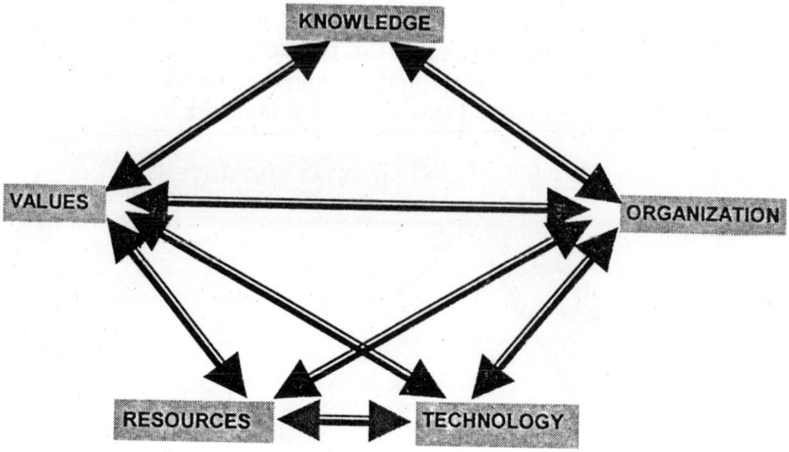

Figure 2.2: A Co-Evolutionary View of Development

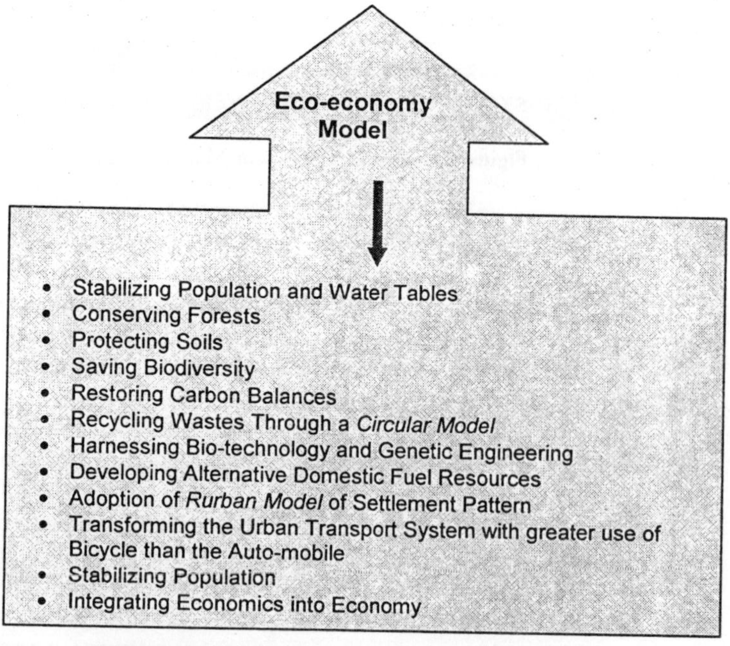

Figure 2.3: Eco-Economy Model (Lester Brown)

Such drastic changes can be brought about only through vigorous advocacy, appeals, exhortations and motivations in which the role of each citizen of the earth is important. The efficacy of this approach has been demonstrated in China by President Mao, who made a clarion call for *Greening the land* by his slogan "*go to Tachai and learn*". As the story goes, people went to Tachai and learnt some lessons in sustainable development.

Awareness building and advocacy are the two major functions of the Bhoovigyan Vikas Foundation. In this context, it is vigorously advocating the celebration of "*Earth-Day*", synchronizing with the equinoxes, when days and nights are equal. Another major advocacy is the building of open *Museums of Earth, Man and Environment*. One such museum is already coming up in Raipur in the state of Chhattisgarh. Undoubtedly a search for a common direction or broad vision is necessary to tackle the challenges of population and environment.

REFERENCES

Brown, Lester R. (2002). *Eco Economy*. Oxford University Press, New Delhi.

Jordan, Carl F. (1998). *Working with Nature-Resource Management for Sustainability*. Harward Academic Foundation.

Malthus, T. R. (1798). *Essay on the Principle of Population*. Johnson, St. Paul Church Yard. London.

McNamara, Robert S. (1984). *The Population Problem: Time Bomb or Myth*. Washington D.C.

Noggard, R. B. (1988). Sustainable Development—A Co-evolutionary View. *Futures*, 20 (6).

3

Policies for our Plundered Planet: Some Pathways to the Future

K. V. SUNDARAM

"Where there is no Vision, the People Perish
So will our Planet, sans Earth Care Vision"

I. INTRODUCTION TO THE FOCAL THEME

A major objective of the Bhoovigyan Vikas Foundation (BVF) is to work towards the common goal of "*Saving our Planet*", through various activities designed to achieve Sustainable Development. It implies a deep concern for the sustainable development and management of our Natural Resources, with conservation and equity considerations. The objectives here are: (i) to widely disseminate and spread the message of Sustainable Development and Living by creating the awareness, need and urgency for protecting the earth from the ravages of human action and (ii) to promote the rational use of natural resources through planned practices of conservation and development. To achieve these objectives, *policy guidance* through carefully conceived Earth Care Policies is necessary, to design a realistic perspective plan for a new Global Society.

This paper is mainly confined to outline the contours of some priority policies and will not go into the details of the current state of environment, the problems emerging from Earth–Man interface and their various consequences, which call for urgent action along several fronts. It is assumed that these are fairly well known and they constitute the starting point for our discussion here.

The paper underlines the fact that humanity's stake in earth protection is enormous during the current century, as the sane earth care policies, necessary in this behalf, have been neglected too often in the past. The basic dilemma here is related to the two-way relationships between development and environment. The twin concerns in this relationship are:

- Environment without development would be life without opportunities.
- Development without adequate environmental protection would inflict serious environmental damage and destroy Earth's delicate natural balance and hasten its very end.

Our Focus

Our country is blessed by the beauty of its vast and varied landscapes and the wealth of its natural resources. But with this blessing, comes the responsibility to ensure their preservation and a healthy environment, in a planet that should be liveable and pleasant. In the new knowledge-based and technology-driven global economy, this can be achieved only by integrating our economic, environmental and social goals and policies with various levels of government, within our country and across the globe. These goals must be achieved *in partnership* with several stake-holders—governments, and the civil society; industries and business; academicians, researchers and technologists—who have the requisite perception, knowledge and expertise to steer the right course of action. In short, *Caring for the Earth* is everyone's business.

Our Strategy

Caring for the Earth falls between too many disciplinary stools and calls for broad and explicit strategies, encompassing a vast and wide spectrum of policies and programmes in numerous sectors of development. The policies and actions needed in this context have to be in the nature of "coordinated efforts", ranging from local to global. They must be conceived as an 'interlinked strategy', mutually supporting each other and aimed at the common goal of sustainable development.

Earth Care Policies: Its Strategic Directions

The issues to be addressed under the term, "Earth Care Policies", will cover a set of carefully conceived proposals for national strategies, plans and processes of development and promoting a value system enshrining the principle of sustainability over generations, of the planet's multitude of natural resources, including Land, Water and the Biospheric environment, with their numerous living species, including the human-kind. This calls for an array of consistent, coherent and properly sequenced actions on multiple fronts by multiple organizations. This is no small task. The present paper will focus on some prismatic and targeted policies relating to those natural resources, which rank high among the *Earth Care Priorities.*

This discussion is organised around the following major heads:

- International Environmental Concerns.
- The Crisis of Multilateralism.
- Environmental Degradation Sustainable National Resources Management.
- Sustainable Water Security.
- Sustainable Agriculture and Food Security.
- Earth Care Policies for Halting Land Degradation.
- Sustainable Forestry and Biodiversity.
- Sustainable Energy and Industry.
- Towards a 'Circular Model' of Materials Economy.
- Sustainable Rural Development Policy.
- Population and Carrying Capacity and Human Settlements Planning.
- Environmental Laws and Economic Instruments for Promoting Sustainable Development.
- The limited coverage of this paper..
- Towards a new paradigm of Caring for Mother Earth.

Several specialists and experts participating in this International Conference have contributed research articles and reported in-depth investigations and case studies carried out by them on particular aspects of the problem. For lack of time, no attempt has been made in this paper to draw upon all these studies and synthesise their findings. This is a task that is proposed to be attempted, as a post-conference exercise. The limited aim of this paper is to *hit at some key issues on Earth Care Policies*, to facilitate our discussions in this conference.

II. INTERNATIONAL ENVIRONMENTAL CONCERNS

Beginning from the Stockholm Conference in 1972, the subject of Earth Care Policies has been addressed in successive global forums, to ensure that human progress should be sustained through sustainable development, 'without bankrupting the resources of future generations'. The Brundtland Commission Report on Environment and Development in 1987, which was followed by the United Nations Conference on "*Environment and Development*" in Rio de Janeiro in 1992 with its agenda 21, and the recent *World Summit* on "*Sustainable Development*" at Johannesburg in September 2002, are significant milestones in the thinking about our "Common Future". They have all sought to advance our knowledge and ideas (at least in a conceptual sense), that would contribute to a style of development, that would be economically desirable, socially equitable and environmentally sustainable. Still in a practical sense, the developing and the developed countries are far from reaching these envisioned goals. On the contrary, there has been mounting evidence to prove that "*our economic policies that have yielded the extraordinary growth in the world economy are the same ones that are destroying its support systems*" (Lester Brown, 2001). The facts and figures thrown up by the World Watch Institute and the Earth Policy Institute are revealing. The complexity of the problems that we face in the beginning of this new century is quite perplexing and the need to act with greater determination, vigour and urgency has become more pressing. In this context, "*the gap between what we need to do to arrest the environmental deterioration of the planet and what we are doing continues to widen. Somehow we have to turn the tide*" (Lester Brown, *2001*).

Since Stockholm, it has become clear that a *holistic, cooperative multilateral and partnership approach* needs to be taken relating to the multifarious aspects of our natural ecosystems—their structure, functions and diversity—with due deference to ethics, values and the promotion of justice and equity. To achieve practical success, we need to recast our programmes and efforts to deal with specific and focused issues. We need to deal not only with the domestic policies of 'local' concerns in regard to air, water pollution and chemical toxicity, but also with the interlinked issues demanding international concerns, like climate change, rising sea level, dying coral reefs and disappearing species. The magnitude of restructuring that is needed to meet these challenges and to advance towards these goals with commitment and dynamism is indeed frightening, as it needs increased national and international efforts for continued action at consensus building and global partnerships for sustainable development. Here there is no way, other than adapting our national and international policies and multinational cooperation strategies to the new challenges and realities of our present millennium. In this respect, humanity, now more than ever, truly stands at a *defining moment* in history.

III. THE CRISIS OF MULTILATERALISM

Among the International Concerns aimed at multilateral efforts, a major one to be addressed is *climate change.* The air we breathe belongs to everyone, but the extent or 'space' to which it can be polluted by emissions like Carbon-dioxide (CO_2) and other green house gases (GHG), is finite and limited. According to one estimate, India is now the World's fifth largest fossil fuel CO_2 emitting country, the emissions having grown at 6 per cent a year since 1950 (Karunakaran, 2002). The only solution by which the emissions could be contained within reasonable limits, will be by sharing this 'space' among all nations on some *equitable basis*. According to the "*strategy of contraction and convergence*", the rich countries should contract their emission levels, to a maximum extent (as they are the greatest polluters), while the poor countries may be allowed, (for the sake of developing their economies) to increase their emissions to a *reasonable level.* The per capita emission and the time for adjusting to safe levels of CO_2 concentration are matters to be negotiated internationally. This strategy will hopefully lead to a just and legally-binding framework for global safety and saving the world from a looming catastrophe. While some developed countries like Denmark, The Netherlands, United Kingdom and Japan have endorsed this principle of '*contraction and convergence*' at the Kyoto discussions, some others like U.S.A., who are the worst polluters, have not agreed to do so.

The World Summit on Sustainable Development at Johannesburg, which took place between August 26 to September 4, 2002, with some 21,000 delegates, including 104 Heads of States and Governments, also could not make much headway in this regard, excepting *urging* the countries who have not signed the Kyoto protocol to do so. Finally, the Johannesburg meet dealt with defining certain deadlines for some specific issues like sanitation and clean water, health, agriculture, environment, renewable energy and poverty. Ultimately, the summit ended as a compromise to accommodate the demands of a powerful few, at the cost of the many poor countries. It could not draw a road map for sustainable development. Despite its failure, multilateralism and global negotiations have to continue with hope against hope.

Following the World Summit, a ten-day *Eighth Conference of the Parties (COP) to the United Nations Framework Convention on Climate Change (UNFCC)* held its deliberations at New Delhi during September-November, 2002, intended to move ahead towards the substantive issues related to the Kyoto Protocol, following its earlier meetings at Bonn (Germany) and Marrakesh (Morocco). In the final Delhi Declaration, the Convention managed to achieve a degree of success on most of the contentious issues, except on the issue of future commitments. A positive progress was in regard to spelling out the concept of *Clean Development Mechanism (CDM)*, by which developed countries can invest in emission-reducing projects in developing countries and thereby earn carbon credits. This summit has certainly paved the way for the ratification of the Kyoto Protocol by 96 countries, which account for a total of 37.4 per cent of the emissions. The process of multilateral negotiations is undoubtedly a complex and long drawn-out one. But ultimately, the dialogue should be able to bridge the differences between the developing and the developed countries, by the sheer compulsions for creating a liveable planet. How this could be brought about is still a moot question. But the only increasing hope in my view, since we cannot afford the luxury pessimism, would be through a *revamped and highly empowered United Nations Organisation*. The tide in the affairs of 'Earth and Man' today beckons us for greater global cooperation and for more coöperative decision-making. Let me also remind the words of our first Patron, the late Shri C. Subramaniam, "there is not much time left for the planet to be saved".

IV. ENVIRONMENTAL DEGRADATION AND SUSTAINABLE NATURAL RESOURCE MANAGEMENT

Developing countries, who are highly dependent on primary production, rely heavily on the efficient and sustainable management of their resource base,—the land, water and the biosphere—which support these activities. A major requirement for formulating comprehensive Earth-Care Policies for halting environmental degradation is the need for a reliable database, which is a constraint in all developing countries. In the absence of such a database, policy decisions are always taken with '*optimal ignorance*'. However, rapid advancement in the techniques of natural resource surveys and developments in Information and Communication Technology (ICT), Biotechnology, Ecotechnology and Geomatics (Spatial Informatics Technology), and their applications in planning through "*fusion of technologies*" has changed the context.

By leveraging these technologies for securing sustainable growth and competitive advantage for value addition, as well as for providing remunerative opportunities for the farmers and empowering them to capture the optimum value for their produce, it would be possible to elevate the Indian farmer and enhance his purchasing power and competitiveness to withstand the challenges of our new century. This must be recognized as the most important among our priorities in framing our Earth Care Policies in agriculture for the future. The policy dimensions in this context would be:

(i) Substantial increase in yields[1] which will enable higher farm incomes;
(ii) Diversification of land use to optimize farm income by better aligning of the produce with the needs of the market, both domestic and international;
(iii) Effective linkages with the food processing industry, to develop and market its value added products in the domestic and international markets; and
(iv) Efficient aggregation mechanisms and logistics to service the requirement for high quality inputs and knowledge at the farm end on the one hand, and to assist in output marketing on the other, making cost of transactions and logistics internationally competitive.

In the above context of enhancing global competitiveness of Indian agriculture, four major problems needed to be addressed are:

(i) Impact of subsidies and incentive structures related to land use planning;

(ii) Neglect of the small producer;
(iii) Degradation of the resource base owing to short-sighted policies resulting in loss of soil resources and desertification; and
(iv) Precision farming for productivity improvement, by adjusting farm practices to match the variations of soil and terrain across time and space, as well as ensuring the timeliness and the method of sowing, irrigation and other agricultural operations.

Earth Care policies in the agricultural sector would also require several other enabling requirements such as:

(i) Redirection of government effort;
(ii) Amendments in the legal framework;
(iii) Rationalisation of fiscal levies;
(iv) Mandatory packaging of food products; and
(v) Bringing together various stake-holders in the food supply chain and creating a "Partnership Web" for mutual benefit.

Sustainable Agriculture and Food Securities

Globalization has brought to the forefront the challenges before Indian agriculture. These include increasing food production and productivity, poverty reduction, hunger elimination, employment generation and environmental protection. Demographically, most of the world's poor live in countries with rapidly growing population, where poverty and population growth are reinforcing each other. The Indian subcontinent, for example, is adding 21 million people a year, the equivalent of another Australia. By mid-century, the population of India, which is already characterized as one of the hungriest among the nations is expected to include another 900 million people. According to present estimates, India has more than 90 million farms of less than 2 ha. Many experts have opined that unless cropland productivity is raised and the yield per crop increased, through multiple-cropping agriculture, as practiced by traditional farming methods, with crop residues being fed to ruminants (to produce more meat and milk) and organic farming, by and large, replacing chemical fertilizers, the hunger problem of India and the persisting land degradation cannot be effectively tackled. Biotechnology's contributions are more likely to come in developing crop varieties that will reduce insecticide use, are more drought-tolerant, or are more salt-tolerant. If genetic engineers can breed salt-tolerant varieties, it would alleviate water shortages. In the other context of improving agricultural production and productivity, perhaps the largest question hanging over the future of Biotechnology is the possible long-term environmental and human health effects of using genetically modified crops. The question often raised is: Will it help us in finding sustainable development paths for sustaining the livelihoods of our rural poor? In our country, this problem could perhaps be effectively tackled through the excellent network of our national and state research centres, development of high yielding seeds, provision of better irrigation, the fostering of location-specific technologies and micro-level planning. Some success stories like the Bio-village experiment in Pondicherry by the M. S. Swaminathan Foundation, the VANRAI model of Mohan Dharia, the process of change effected in Ralegan Siddhi by Anna Hazare, the Bharatiya Agro-Industrial Foundation's (BAIF) model for sustainable livelihoods, the work of Ramakrishana Mission in rural areas are some examples. In all these cases, economic and social change through voluntary action by civil society participation has been a potent, combative and constructive force, laying the groundwork for building a better society. The biovillage concept of the Swaminathan Foundation addresses the ecological foundations for sustainable agriculture and providing sustainable livelihood opportunities. It is a pro-nature, pro-poor, pro-women development paradigm to sustainable agriculture and rural development. Bhoovigyan Vikas Foundation's advocacy of a *model of auto-kinetic development* (Figure 3.1) with its underlying concepts of enhancing "recycling ratios" in the village, establishing "integrated economic circuits" in village clusters and the promotion of agro-industries in accordance with the principle of *subsidiarity*, conforming to Christaller's concept of hierarchical planning of economic activities and social services is a further elaboration of these models, leading to holistic sustainable development, technological empowerment and social mobilization, operated on a 'village cluster planning' mode.

Some Postulates and Policy Dimensions

In our quest for sustainable agriculture, the following postulates and policy dimensions must be recognized:

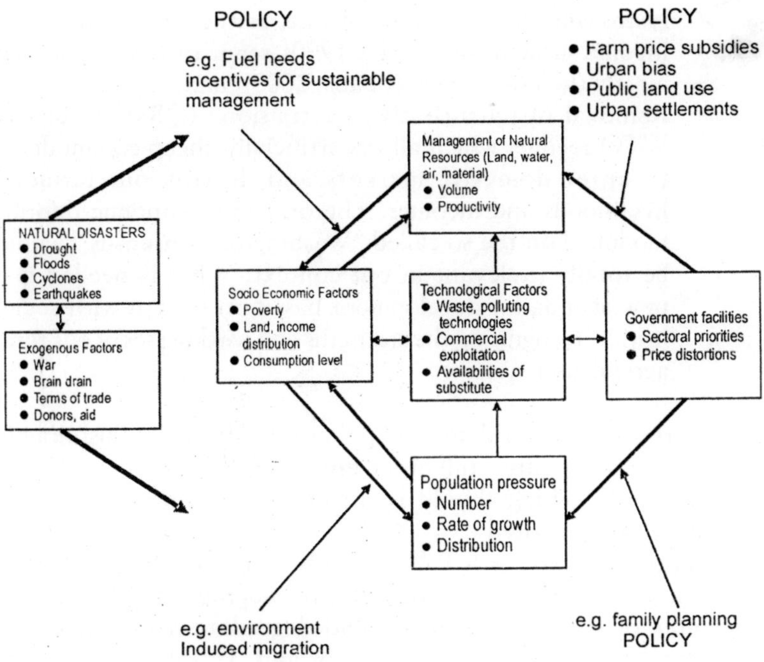

Source: Adapted from ADB, February 1990.

Fig. 3.1: The Linkages between Demographic and Natural Resource Issues

- Ensuring food security;
- Increased access to land, water, fertilizers, pesticides, markets, infrastructure and other requirements;
- Organising the small farmers and the rural poor;
- Bureaucratic and Extension reorientation;
- Working with community-based organisations and para professionals;
- Adopting the farming systems approach;
- Re-orienting the activities of small farmers' production systems, in order to assist them to mobilize capital through joint credit and other means to enable them to enter the market economy;
- Community management of common property resources;
- Encouraging non-farm activities to reduce poverty-levels in rural areas both by way of contributing to the GDP as well as by increasing the employment potential to absorb surplus labour in the agricultural sector;

- Narrowing the rural-urban divide, as the benefits of better income growth during the 1990s seem to have benefited mostly the urban areas (Bhalla, 2000); and
- Removal of Quantitative Restrictions (QRs) under the WTO regime, which allows artificially cheap commodities to enter domestic markets and destroy our farmers' livelihoods and incomes. The on-going economic reforms modelled on the so called "Washington consensus", would be totally irrelevant in our context. What is needed is a new agenda for development based on growth with social justice, recognizing the strengths and weaknesses of our rural agricultural system.

All the above postulates and policy dimensions of Sustainable Agriculture constitute a tall order, entailing a big bundle of "Earth Care Policies", which should be unambiguously formulated and rigorously implemented.

Another major Earth Policy consideration in the agriculture sector is meeting the challenge of land degradation. While the estimates of the magnitude of land degradation in India vary considerably, (ranging between 23% and 32.75% of the total geographical area of the country[2]), it is definitely a challenging problem demanding urgent attention. According to Ravindra Babu, "*Even if one* million hectares of land are successfully treated every year, it would still take more than 50 years to recharge all the affected areas". With land degradation are the associated continuous geomorphic and anthropogenic processes like soil erosion, reduction in forest cover, over-exploitation of the ground water, changes in the cropping systems, loss of crop diversity and a rapidly growing population which impinge on the land, affect food security and threaten the livelihoods of millions of people.

Land degradation, particularly in dryland regions, is characterized by increasing conflicts between different users for control over farmland, water and pastures. More successful instances of combating land degradation comprise of small scale programmes, based on simple technologies, developed with the full participation of local people and providing fairly immediate benefits to them. Earth Care policies must therefore confer special rights and responsibilities to manage natural resources by "*local user groups*".

In dryland regions again, food security and water security are often concomitant problems. These have to be addressed through both government as well as by local organizational efforts. Food security measures by government would include the implementation of efficient public distribution system and 'food for work' and other kinds of employment programmes, while community efforts would be by building "Food Banks" or "grain golas" and other cooperative methods. Water security measures would also require the attention by both government and local efforts. Lessons have to be drawn from small scale, locally appropriate water security projects initiated by NGOs and rain water-harvesting measures practiced in local areas by people, based on indigenous knowledge and traditional practices.

V. SUSTAINABLE WATER SECURITY

In the above context of agricultural planning, developing and managing our water resources on an integrated basis, in order to meet the present and future needs for agricultural production on a sustainable basis, bearing in mind the competing demands for limited water resources, the rational use of water and an efficient water management strategy have become critical issues. It is not too much to assert that high priority should be accorded in our policies to the adoption of urgent measures for soil and water conservation, within the framework of integrated land and water management, with due regard to long-term as well as short-term productivity.

The importance of water in Earth Care policies for sustainable development arises mainly from the problems of water scarcity, resulting from uncoordinated and badly planned land and water development as well as water quality, and degradation due to agricultural, industrial and municipal water pollution. The related issues on sustainable water resources development are the poverty, environmental degradation and population growth.

The following five priority areas for action are indicated for the development of water for sustainable agriculture (FAO):

(i) Efficient water use;
(ii) Waterlogging, salinity and drainage;
(iii) Water quality management;

(iv) Small-scale water programmes; and
(v) Scarce water resources management.

While the broad measures for alleviating the problems of scarce water management are generally known, the areas of conflict arise in the "frictional losses" in implementation, as government administration in federal countries are characterized by marked regional interests, the latest Cauvery issue in our country being an example. This is not the place for further analysis of this issue, as it is mixed up with legal issues and the approach of political economy. We shall, however, confine ourselves to some policy dimensions involved in water as a critical factor of life-sustaining resource—its quality, access, supply, management, treatment and disposal, all of which have a direct bearing on 'nutritional security'. In the definite movement in our changing times towards an ecology-oriented river management, re-naturalisation and conservation as well as progress towards a concept of ecologically and environmentally compatible form of drainage, the triangular relationship between the terms "*Water, Protection and Human beings*" stand out as one central concern in our Earth-Care Policy measures. More specifically, river water management, water supply both for drinking and irrigation and waste water treatment and disposal constitute the cardinal dimensions of a sound water-care policy. This will require main attention to the following earth care policies for sustainable water management:

(i) Exploration and data collection for assessing water availability from the point of sustainable yield;
(ii) Thrust on carrying out objective environmental impact assessment in regard to new projects and especially for evaluating the effects of deforestation, as they affect water projects and for enhancing experience and technology transfer among developing countries;
(iii) Instituting strict water use and conservation rules with legislative requirements and ensuring monitoring, evaluation and feedback systems on a long term basis;
(iv) Developing and implementing appropriate land use and environmental programmes in upper catchments; and
(v) Identifying and implementing appropriate environmental protection measures across the entire watersheds.

The normal practice of construction of water harvesting structures, such as check dams etc. are costly in developing countries. Instead, promoting traditional water harvesting measures, e.g. farm ponds at strategic locations should be the preferred solution. In this context, the major thrust of the policy should be on enhancing the role of local communities in implementing and managing water developments.

In arid semi-arid and drought-prone areas, besides the policy measures enunciated above, the emphasis should be on the optimization of water resources exploitation and raising the level of reliability of supplies through conjunctive use of surface and ground water and ensuring proper functioning of existing structures, conveyance systems and other assets through adequate maintenance and operation. Basic to all these measures would be by promoting beneficiary's participation in all aspects of water planning and management through Water User's Associations and Micro-level planning.

Since the introduction of the Eighth Five Year Plan in India (1992-1997), focus is shifting towards the watershed development and management programmes at the micro-level. The watershed, which is organically linked to the natural drainage system and hierarchically integrated from lower to higher order drainage networks, constitutes an ideal unit for conservation, upgradation and utilization of water resources and ensuring restoration of ecological balance in rainfed areas. Earth Care Policies in relation to watershed planning and development should address issues such as land owner's rights, conflicting issues like individual *versus* group ownership of resources and guidelines for adapting new or changing land use pattern. Such policies should focus on raising family income, increasing farm productivity, diversifying economic activities to provide better opportunity for earning livelihood. The emphasis should be on decentralization of decision making, so that the development plan would flow from user groups and associations, self-help groups and local people.

VI. NEEDED A MULTI-PRONGED STRATEGY

To tackle the Sustainable Development of Agriculture, we need multi-pronged strategies and policies, not merely in terms of economic and technology innovations, but by adopting various other non-conventional methods to address the issues of food security, nutrition adequacy, rural income generation, provision of employment opportunity and removal of poverty. The major components of such a strategy, which must find a place in our Earth Policies are:

(i) An improved R & D system with a "*Knowledge Information and Communication Systems Centre*", installed in each village to bring extension services and knowledge dissemination concerned with developing techniques that would contribute to sustainable agriculture systems and providing easy access to all farmers—both knowledge and services;

(ii) Effectively operationalising the new concepts of participatory democracy, effective local governance and participation of the civil society and NGOs and new institutional arrangements to face the challenges of quality and competitiveness in this era of globalization and liberalisation;

(iii) Conceiving and devising new organisational receiving mechanisms and other changes to tackle the problem of our fragmented production base, operated by a large number of small farmers, with limited reach for modern technologies to overcome their biggest handicaps for competing in the export markets with the necessary capital and market intelligence. One of the methods would be to bring our *Corporate houses* into agri-business jointly with the producers (Farmers);

(iv) Promoting increased '*localism*' in the planning process of the agricultural sector and ensuring the greater involvement of local groups and the private sector to play a more active and instrumental role in agricultural development;

(v) Bringing about *effective* partnerships for synergised action among various stakeholders, for translating a strategy of efficient resource use and management. The partners would be individual farmers, interests groups, community-based organisations, captains of Industry and Trade, Panchayati Raj institutions and local leaders;

(vi) Adopting a research thrust on managing the factors of production and adapting individual technologies to suite location-specific peculiarities;

(vii) Changing our *extension approach* from a fragmented message-based process to a broad-based '*farm management approach*'. The shift should be from a '*contact-farmer approach*' to a '*group approach*', which will promote social cohesion and collective action, essential for marketing, conservation and use of natural resources;

(viii) At the level of public policies, investments and services, the adaption of a '*convergence approach*', by which all public investments in infrastructure—irrigation, power, transport, communication, storage etc. and provision of services—research, development of infrastructure and rural credit etc. can be brought together in mutually supportive and sequent actions to sustain the emerging market-led agricultural growth;

(ix) The widening of the scope of non-farm employment by attracting capital flow into the rural sector through an "*auto-kinetic model*" of rural development. This model will increase the value-added income of the village and establish a mutually supporting hierarchical mode of rural development, based on the Gandhian model, incorporating the "*subsidiarity principle*". In this way, an integrated approach of convergence of markets, jobs and investments would follow and rural-urban integration would be secured. Through '*micropolises*' and rural service centres serving as nodal points for services and employment opportunities, a "*Rurban Model of settlement pattern*" would emerge;

(x) To enhance corporate capital investments in agriculture, restrictive land ceiling laws on dry lands, particularly waste lands must be revised.

VII. SUSTAINABLE FORESTRY AND BIO-DIVERSITY

Since forests are important for climate regulation, flood control, soil conservation, water regulation, nutrient storage and recycling, fertilization, seed dispersal and recreation, which are all a basic part of any economy's support systems, policies for protecting forest products and services assume great importance in our strategies of caring for the Earth. Sustainable forestry practices can be achieved through:

(i) Maintaining the harvest of all products at sustainable levels by careful *control of harnessing* levels, timing and frequency;

(ii) Maintaining essential *eco-system processes*, e.g. by retaining continuous vegetation cover, returning nutrients to the soil etc.
(iii) Maintaining *biological diversity* at eco-system, species and gene levels;
(iv) Satisfying *the needs of the people* living in and around the forests; and
(v) Ensuring economic sustainability on the part of the forest user and on the part of governments.[3]

In the above context, the earth care policy framework for conservation of sustainable development of forests should include the following major ingredients:

(i) Halt inappropriate incentives like the subsidies, logging concession terms, tenurial policies, pricing policies and so on;
(ii) Consolidate the existing settlements in the forest and on its fringe. This would mean providing appropriate incentives like the infrastructure, tenurial and financial security, technology, knowledge and resources;
(iii) Integrate forestry into rural development by paying attention to land and tree tenure, sustaining and developing traditional forestry knowledge, integrating forestry and agriculture and encouraging participation in multipurpose forestry/agro-forestry. This implies initiating joint forest management programmes with the people, who depend on the forest for their livelihood;
(iv) Set up a clear forest ownership structure with a framework of regulations and incentives;
(v) Coordinate the authorities responsible for forest functions, notably, forestry, water, and agriculture to achieve convergent approaches to the management of forests; and
(vi) Ensure participation of local people in all policy planning that affects forests and in forest management and use.

Sustainable forestry has to be vigorously promoted, which means not merely in making arrangement for the sustainable yield of timber, but also encouraging the capacity of the forest to supply both *products and services* sustainably. The measures here will include banning or putting restrictions for logging, preventing export of forest products, sustainable forest management, reducing quantity of wood use to manufacture paper and reducing use of wood as fuel. The long term policy to the fuel problem is to develop alternative sources of energy for cooking based on wind, solar or geo-thermal energy and to develop indigenous sources of renewable energy, whatever they may be.

Biological diversity has assumed great importance because of the destruction of eco-systems and species extinctions, all of which cause irreversible losses. A broad range of measures are needed to preserve biological diversity. The policies should be both to promote development and at the same time relieve excessive pressure on natural resources. The following measures are needed:

(i) Measures that raise yields in agriculture and thus reduce the need for area expansion.
(ii) Policies that increase non-farm employment opportunities such as efficient development policies for trade, agriculture and industry.
(iii) Sustainable forestry practices that remove subsidies for logging and other activities that cause deforestation.
(iv) Encouraging programmes designed to capture the nature of biological diversity with its diverse range of food, medicines, and other products critical for the local population, thus promoting a development of options for the sustainable use of resources.
(v) Programmes to add value to biological resources (e.g. genetic prospecting).
(vi) Development of eco-tourism, which can improve, beautify and manage the landscape of the countryside, ensure an intelligent use of mankind's creative ability without destroying the sensitive balance of the ecosystem or destroying areas of natural beauty.

VIII. SUSTAINABLE ENERGY AND INDUSTRY

Biomass and fossil fuels have continued to be the largest sources of energy in developing countries so far. Hydro-electric power has also been an important source of energy, though it is only next to coal-fired power. With carbon emissions being comparatively lower in the developing countries, pollution abatement technologies are not widely deployed in the developing countries. But the scenario has now changed, with the emissions of pollutants rapidly increasing and the threat of greenhouse warming, compelling the restriction on the use of fossil fuels. Hence the use of alternate sources of energy is now necessary to reduce emission of CO_2 per unit of output. The options available are renewable energy, like the solar energy, biomass, geo-thermal energy, hydro-power, wind and nuclear energy. While taking policy decisions on alternative sources of energy, governments have no doubt to take into account energy production costs and energy-related pollution. Thus the major considerations for promoting sustainable energy in the developing countries are:

(i) Cost effectiveness as well as increasing the efficiency of energy use to reduce its polluting effects;
(ii) Promoting the use of renewable and alternative sources of energies in households and vehicle fuels;
(iii) Promoting Industrial development in such manner, that the emissions from their activities are such that they can be readily assimilated by the environment, without causing serious environmental problems; and
(iv) All available technological innovations—pre-combustion, combustion and post-combustion technologies—must be used to control emissions and increase energy efficiency. These must be accompanied by measures to exercise strict environmental controls through enforcement mechanisms and progress monitoring.

Earth Care policies designed for each country context must keep the above considerations in view and aim at the following measures:

(i) Formulate national action programmes to promote and support re-afforestation and forest regeneration to achieve sustained provision of adequate biomass energy to meet the needs of the rural and urban low-income groups. In this context, improved biomass stoves should also be promoted and subsidies provided for the development of such stoves.
(ii) In the new century, the wind power is being advocated as the foundation of the new energy economy, which through electrolysis can supply both electricity and hydrogen for fuel.

Similarly turning sunlight into electricity through solar cells and tapping the Earth's heat as Geothermal energy provide other sources of alternative renewable energy resources, which are bound to accelerate the transition to a new energy era during the present century. According to futurologists, in this transition, natural gas would overtake as the *transition fuel* during the first quarter of this century.

(iii) In the new *eco-economy* visualized by Lester Brown, wind farms would replace coal mines, hydrogen-powered fuel cells would replace internal combustion engines and the solar energy with solar cells would play the most important roles, in replacing the old energy economy. One of the characteristics of the new energy economy will be that it will rely much more on *decentralized small scale power sources*, rather than on a few large centralized systems.

IX. TOWARDS A CIRCULAR MODEL OF MATERIALS ECONOMY

Managing the Materials Economy

In our modern lifestyle in this material age, we have tended to generate a huge volume of garbage, wastes, metals, throwaway products and toxic chemicals, which damage the environment. Much of this material could be rendered regenerative, by processes of recycling. Scientists call this as "*redesigning a new materials economy*"; replacing the current *linear* "*flow-through model*" with a "*circular model*", that emulates nature, *one that closes the 'loop'*. In the mining industries for example, the scrap can be reused through recycling. Similarly water can also be recycled. Redesigning also includes discouraging products that are polluting and non-disposable through products that are easily bio-degradable and disposable. The classic example is paper *versus* plastic. In our agenda of Earth Care Policies, *the circular model* of designing a new materials economy should occupy an important place. Regulating hazardous wastes in line with the 'Polluter Pays principle' is yet another method to control hazardous wastes from industrial sources. An autonomous "*Industrial Environment Fund Organisation*" could be established to levy *waste charges* from each industry, to establish and operate "*central treatment and disposal*" facilities, as done in Thailand. This is a method of minimizing pollution control costs and encouraging efficiency in industrial production processes. It can go a long way in preserving our natural resources.

X. SUSTAINABLE RURAL DEVELOPMENT POLICIES

Our rural landscapes, once ruled by virgin nature and viewed as a "terrestrial paradise", and a natural heritage, as well as a strong and safe retreat for man's recreation and rejuvenation is getting transformed into an "exhausted mine" owing to a "pillage economy".

This pillage economy works on a short term "profit logic" of intermediate struggle for survival, with disastrous assaults on our ecologic environment. In the area of agriculture, this short term approach is reflected in the term *agribusiness*. It thrives on a myopic view of a world of limitless natural resources and reliance and faith on substitute technologies for the vanishing natural resource capital. With the introduction of green revolution in the annual cycle of the three food crops of wheat, maize and sorghum, against a decennial need for soil replacement and recuperation, it has led to several unintended results and environmental costs. It is totally in contrast with our secular, long-run traditional agriculture systems, evolved on the basis of traditional ecological knowledge and intuitive experiences of societies.

The contrast between these two techniques of food production has led to environmental problems and raised the question: how to manage the ecologic environment that give the people their sustenance? Thus a commercial approach rooted in the greed for speculative gain and the inability of the traditional farmer to adjust to a technologically oriented society is the heart of our rural problem. This is not a simple matter, because of the complexities of the interactions that determine the dynamics of the natural environment. This represents one aspect of the optimization and utilization of land and water resources in the framework of a change in land use, which the modern economy tries to bring about in the form of agri-business. Obviously, effective policies of Earth Care, oriented to the degree of tolerance of the natural environment to competing uses and conflicting value systems are the need of our times. The 'How' of it is the tricky issue, demanding a development policy which requires a mutual process of adaptation of natural resources, technology and of man himself. This is not a simple matter because of the complexities of the interactions that determine the dynamics of the natural environment. The old saying "you can't see the wood for the trees" is very apt to the current situation. What is needed in this context is to understand the complexity of nature and the web of interactions that manifest themselves in its dynamics.

How to develop our rural environment rationally, and to manage it efficiently, that is by assuring the non-degradation of its resources are the imperatives that impose themselves all the more forcibly in this new century, as the world situation is characterized by:

(i) an ungovernable increase in population;
(ii) a vast majority of these people (between 2/3 and 3/4) who are poorly or insufficiently nourished to grow into healthy individuals; and
(iii) an industry, with its quest for profit causing immeasurable deleterious effects on the ecologic environment.

With the increasing disparity between rich and the poor and the patterns of production, consumption and distribution followed in several developing countries being ecologically unfriendly, economically biased and socially unjust, a new paradigm of development is called for. The situation obviously is serious and tragic. Only the force of public opinion can avert a catastrophe.

In the above context, some important policies that will ensure *Caring for the Earth*, may be outlined. They will include:

(i) The removal of those government policies that exacerbate environmental problems, such as subsidizing the extraction of scarce resources;
(ii) Clarifying property rights, since insecure property rights over land and other natural resources have led people to misuse them or under-invest in their improvement;
(iii) Encouraging flexible types of environmental regulations that rely on economic incentives to encourage compliance. Simpler methods of environmental regulations, such as imposing charges on polluting inputs, rather than on the pollution itself or providing deposit refund systems to induce producers and consumers of polluting products to return waste for recycling or treatment and safe disposal.

To improve the capacity to implement the above mentioned changes, governments must shift their strategies from top-down management to one focused on strategic intervention. A decentralized approach instead of a huge regulatory bureaucracy would ensure that environmental impact is considered in all areas of policy making. Central Government should devolve responsibility for environmental management to appropriate local and regional levels, and enter into partnerships with the private sector through non-governmental organizations (NGOs) and local communities. Determining the appropriate level of government for environmental management can be determined by considering at which level can effective measures be undertaken. For integrated river basins or watershed management, for instance, institutional arrangements may have to straddle the jurisdiction of several States.

Civil society needs to be involved more intensively. Environmental Impact Assessments should be done in a way that stakeholders can participate at the project level. Thus in the new model of Earth Care Policies, greater decentralisation and broader participation are visualized as the essential prerequisites to work successfully and to move forward in the direction of sustainable development.

Since one view is that poverty is the greatest polluter, poverty reduction has to be an overarching goal, with its strategic concerns, such as human development, sound environmental management, improvement of women's status and fostering economic growth for increasing employment opportunities.

The three pillars of the new poverty reduction strategy are: fostering sustainable economic growth that benefits the poor, promoting social development and encouraging good governance. These strategies have to be prioritized and brought into comprehensive mutual-specific action plans.

In the poverty reduction and environmental care strategy, increasing the access of the poor to essential services, including electricity, water supply and sanitation and most prominently to basic education and primary health care should be addressed to increase the poor's chance of earning income and providing social safety nets.

In all the above policy changes proposed, governments must obviously play a more active role in providing the critical public goods, such as agricultural research, extension services and rural infrastructure. Some aspects of development like the provision of micro credit to the rural poor, for instance could be promoted as *self-help* measures. The Gramean Bank's business strategy based on the voluntary formation of small groups is an example to mobilize the social capital in closely-knit communities in the developing countries.

For transformational change in rural India and meeting the challenge of increasing growth, a threefold strategy of (a) leveraging information technology for rural transformation,(b) leveraging biotechnology for sustainable growth, and (c) linking farmers to remunerative markets would lead to rural partnerships to enhance competitiveness. The first one for instance, will bring about knowledge-based empowerment of the Indian farmer and will also enable creating infrastructure for agri-supply chain management and rural distribution. The second strategy, that of leveraging biotechnology will support large scale farming and social forestry, together with watershed management to create additional avenues of income to the rural poor. The third strategy of linking farmers to remunerative markets will provide the fillip to the food processing industry and lead to value addition to the farmers, by opening up remunerative opportunities to them.

To move in the above directions, the corporate sector companies should develop a "*commitment beyond the market*". The Indian Tobacco company's strategy followed on these lines by establishing ***'e-choupal'*** in a number of villages has led to transformational change in rural India. This example is worth emulating.

As the population continues to grow, in the next 50 years, demands for food, fuel and fibre will rise enormously. In order to meet these demands, more intensive and extensive exploitation of many natural resources will be required. The stress will be more on agricultural land, forests, water and fisheries. It is in this context that sensible resource management is called for. Food demand is expected to double by 2050. All these things mean that an increase in expenditure for agricultural research and development is very important. Yet, expenditures for agricultural research are stagnating in many countries.

All the above changes require carefully conceived and prioritized policies in several areas.

XI. POPULATION AND CARRYING CAPACITY AND HUMAN SETTLEMENTS PLANNING

According to UN projections, the world population will grow from 6.1 billion in 2000 to 9.3 billion in 2050. The bulk of this increase is expected to be in the developing countries. The medium level projections of the UN indicate that the population in the developing world will increase from 4.9 billion in 2000 to 8.1. billion in 2050, i.e. more than 2.5 times. This would lead almost to the collapse of our ecosystems in many countries. In the developing countries, the linkages between demographic and natural resource issues are dynamic and complex (see Figure 3.1). Population growth exacerbates the pressure on the resource base at any given level of technological development. This leads to further deterioration of the earth's environment and in the socio-economic condition of the people, with increasing poverty levels and rapid population growth. This is particularly characteristic of the developing world. Another important trend is that this rapid population growth in the developing world is becoming more and more concentrated in cities, which are already facing the problems of increased influx of rural population, leading to congestion, pollution, growth of slums and squatter settlements.

In recent years, studies have not been lacking to suggest measures for containing the population growth to match the carrying capacity of the earth. But government policies and measures have not been equal to the task and implementation of programmes in this direction has been woefully lacking in practice.

Some policies suggested for the developing countries are:

(i) Filling the family planning gap and increasing the access to family planning services, by providing a full complement of such services, including reproductive health.
(ii) Removing the physical and social barriers that prevent women from using family planning services.
(iii) Creating social conditions that will overcome inhibitions and lead to fertility decline and smaller families.
(iv) Promoting female education in a substantial measure, which has the potential of reducing fertility by roughly 10 per cent (Summers, 1993).
(v) Reducing the average family size to 2 children or less.
(vi) Changing people's attitudes about reproductive health, gender equity, family size and environmental protection through informal education methods, e.g. soap operas (serialized dramas) and sitcoms (situation comedies) aired on radio and television. The impacts of such measures have

been studied and results seem encouraging. Pancha Polston (2000) has reported that soap operas in Mexico, which continued over a decade, has led to a fall in birth rate by 34 per cent and in promoting progress in achieving replacement-level fertility. According to Population Reference Bureau, (2001), 54 countries have reduced average family size to 2 children or less. Among these, Japan and China in Asia figure prominently. The UN's low assumption projection for 2050 (base year 1950) assumes that the entire world is likely to quickly move below replacement-level fertility to 1.7 children per couple.

In our Earth Care Policies for matching population with carrying capacity, developing countries should accord the highest priority to population issues. Family planning, reproductive care, health, environmental and human development programmes should be brought together and designed as a conjunctive programme. Educating girls, enabling women to earn cash incomes and participate fully in decision-making and vigorously pursuing family planning programmes should be conceived as a package of inter-related programmes in rural development. To bring about such close coordination, all those departments, which are now functioning independently or in different ministries should be brought together into one "Population Ministry". This would constitute an important institutional reform in government.

Human Settlement Planning

Human settlement planning, including halting urban sprawl and promoting manageable cities is yet another dimension of Earth Care Policies. The United Nation's projections indicate that by 2050, more than two-thirds of the world's population will be living in cities. Since cities exert a heavy burden on the earth's eco-systems, earth care policies should promote a "*Rurban Model*" of settlement planning. This would mean reducing urban sprawl and encouraging growth of intermediate centres of optimum population, which will reduce crimes, lessen rural urban migration, create new jobs in small towns and provide all the needed urban services to the rural areas, such as education, health and access to goods and services.

In order to tackle the problem of overgrown metropolises and reducing the population in them, the automobile-concentrated urban transport system is discouraged in most policy statements, and instead *the bicycle human friendly transport system* is being advocated. To solve the problem of parking place in cities, parking taxes, declaring car-free areas in cities, promoting sharing by commuters are being tried in several cities. However, in the ultimate analysis, the best policy measure would be promoting a "*micropolis-centred human settlement system*", which is going to be a challenging task to the countries, as it would demand a good deal of dismantling the present systems and restructuring the existing style of the man-made pattern of settlements.

XII. ENVIRONMENTAL LAWS AND ECONOMIC INSTRUMENTS FOR PROMOTING SUSTAINABLE DEVELOPMENT

There is a general feeling that the costs of protecting and improving the environment would be very huge. The World Development Report 1992 on "*Development and Environment*", however, has categorically stated that "the costs of addressing the main environmental priorities are *affordable*—some because of the improvements in economic efficiency they bring, others because of their environmental benefits". According to them, most investments will ultimately pay for themselves. At the same time, they have also mentioned that increased international support will be essential to the developing countries. More recent studies have identified some policy instruments for building an environmentally sustainable economy.

Environmental Laws

These constitute an important set of tools to give effect to policies needed for sustainability. Effective policies and comprehensive legal frameworks that safeguard human rights, the interests of future generations and the productivity and diversity of the Earth are important elements in providing a national framework for integrating development and conservation. Environmental law requires standards of social behaviour based on scientific understanding and a clear analysis of social goals for guiding and leading communities to live within the capacity of the Earth.

Economic Instruments

These include:

- Imposing Resource taxes for limiting the demand, as an incentive for sustainable behaviour. They could replace existing taxes; or the money they raise could be returned to the tax-payer as subsidies for more sustainable changes and progress.
- Supplementing Resource taxes by charges that apply "*the user pays*" principle, e.g. pollution charges, tourism dews for protected areas and water charges for irrigation.
- Removal of subsidies that encourage resource depletion and environmental degradation.
- Deposit/Refund Schemes and Performance Bonds for ensuring that the economic costs of environmental protection reflects the true social cost.

Priority Actions

Earth Care Policies demand several *priority actions* that would require acceptance by every nation, as its share of responsibility and commitment, and act upon it as much as its means would permit. Strengthening International law and existing International agreements to conserve life-support systems and biological diversity are part of these priority actions to help achieve global sustainability. The United Nations system should function as an effective force for global sustainability. A few summits which only end up as verbal exchanges and diluted recommendations will not do. UNEP, FAO, UNIDO, WHO, UNESCO, UNICEF and all other International organizations must take much more effective follow-up actions.

Restructuring the Global Economy

Restructuring the Global Economy to make it compatible with the earth's eco-system, so that economic progress can continue along with ecologically sustainable development is an important aspect of Earth Care Policies. Lester Brown has argued that Economy is part of the Environment, and not the other way. He has stressed the following points:

- Ecological costs of various economic activities should be incorporated into the market price of a product or service in the form of a tax (Environmental tax).
- Other policy instruments that can be used to restructure the economy would include fiscal policy, government regulations, eco-labelling, tax shifting, subsidy shifting and tradeable permits.

Thus it is argued that restructuring the tax system is the key to eliminate the crippling economic distortions. According to Brown, "if they use fiscal policy to encourage environmentally constructive activities and to discourage destructive ones, we can steer the economy in a sustainable direction". Tax shifting, i.e. changing the composition of taxes but not their level is another means used by many countries in Europe. It means reducing income-taxes for refraining from destructive activities. Lester Brown emphasizes that "if the world is to restructure the economy before environmental destruction leads to economic decline, tax restructuring should be the centre of the effort".

Subsidy shifting is another method to encourage environmentally constructive activities through financing them by shifting funds from environmentally destructive subsidies.

Eco-labelling is now used in many sectors of the economy, including energy-efficient household appliances, forest products from sustainably managed forests, fishery products from sustainably managed fisheries and green electricity from environment-friendly renewable resources.

The whole argument advanced by Lester Brown is based on the fact that the ecological costs must be incorporated into the market price of a product or service in the form of a tax or other economic instruments. While in principle, this will be a good strategy, the worrying aspect is whether in practice, decision makers would be able to understand the ecological consequences of their decisions and give effect to those decisions. Brown himself is aware of the problems and says, "*it simply may not be possible*".

When all is said and done, one cannot but emphasise the advantage of increasing *public awareness* on environmental issues and *political will* to enforce environmental laws and regulations, as the more effective solutions from the perspective of the developing countries. If these could be effectively implemented, the Earth's environment could be preserved for future generations to enjoy.

XIII. THE LIMITED COVERAGE OF THIS PAPER

Our discussions in the previous pages show that caring for the Earth is not an easy task and the various problems identified cannot be solved overnight by some new vision. A part of the solution could be by Governments acting with strong will and International organizations also strengthening the pressures on Nation States. But this alone are not enough. The civil society on the ground and the attitudes and practices of individuals have also to respond favourably. A concerted effort is needed to inform and educate people about the need for Sustainable Development and Sustainable Life Styles. Thus moulding the public opinion is an important aspect of the strategy.

In this paper, we have outlined a number of policies for promoting a healthy Earth. Since Earth Care Policies encompass every sector of man's activity, all these could not be covered in a short paper like this. We have touched on some important sectors only. An illustrative scheme is outlined in Figure 3.2, which indicates the principles and commitments needed for effective policies for a healthy Earth, the action strategies necessary as well as some parameters of Earth Care Policies—the Policy orientations, the actions needed for creating a scientific database as a decision-support system (IT strategies) and some broad indications of certain interventions required in the form of fiscal incentives and disincentives, laws and regulations for protection, conservation and abatement of environmental problems, the need for community empowerment on various issues and the Public Sector interventions needed for tackling the critical issues etc.

Towards a New Paradigm of Caring for Mother Earth

The major principles, strategies and actions set out here are those meant to conserve the vitality and diversity of the Earth. They are nothing new and have also been emphasized in many National and International documents. They are meant to be a guide and an agenda for debate, rather than the prescription.

The principal ingredients for a new paradigm for Sustainable Development are as follows:

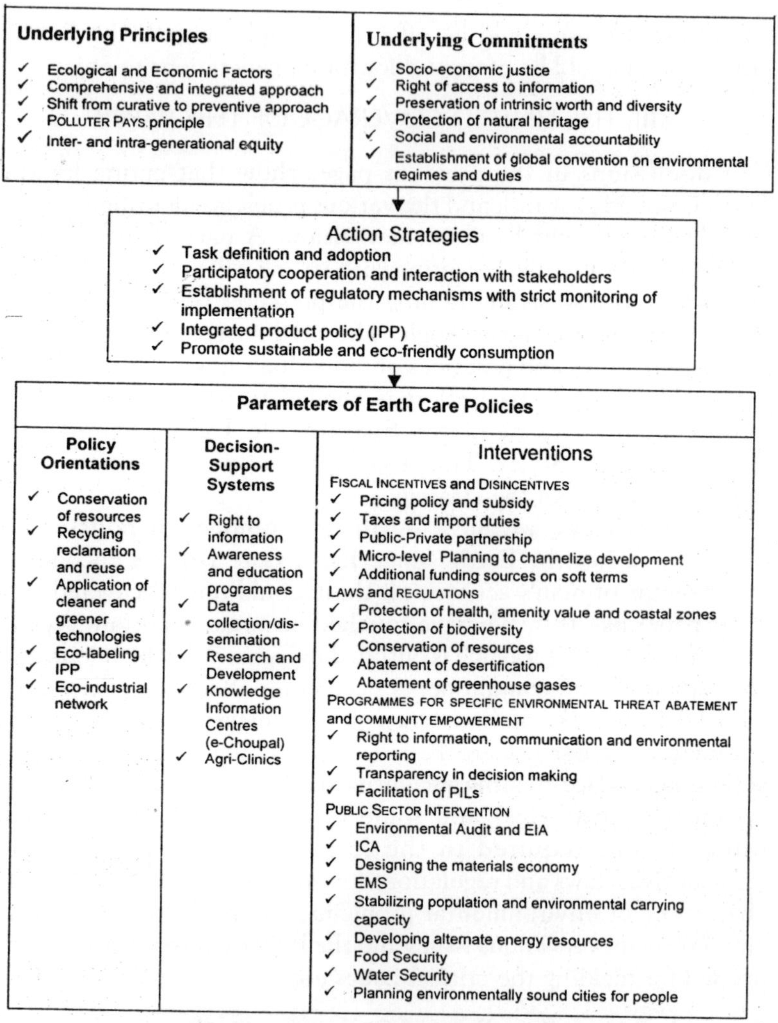

Fig. 3.2: Policies for a Healthy Earth

(i) Minimizing the depletion of non-renewable resources. Where possible, their "*life*" can be extended, for example, by recycling, by substitutes etc.

(ii) Keeping the earth's eco-systems within its carrying capacity, so as to bring human numbers and life styles into balance with nature's capacity, with the use of benign modern technologies to enhance the capacity by careful management.

(iii) Conserving the earth's vitality and diversity by preserving its life-support systems, its bio-diversity and ensuring that uses of renewable resources are sustainable.

(iv) Changing the personal attitudes and practices, by re-examining indigenous wisdom, ethics and values of the society and supporting any new practices, only if such policies and actions are absolutely needed for the survival of the planet.

(v) Enabling local communities to care for their own environments, by encouraging socially valuable actions through their own institutions and community-based organizations.

(vi) Providing a national and regional framework for integrating development and conservation.

- *The national framework* should deal with issues emanating from globalisation, liberalization and world trade regimes, keeping in view the national interests.
- The *regional framework* for integrating development and conservation must take into account the interactions with land, air, water, organisms and human activities. They must relate activities and actions to the scale of the regions—micro, meso and macro units.

(vii) Providing universal primary education for all children and reducing illiteracy.

- Ensuring the total enrolment and attendance at primary and secondary levels.
- Narrowing the gap in female and male literacy and bringing them at par.

(viii) Enhancing security against natural disasters and social strife by evolving proactive as well as risk-reducing measures in disaster-prone areas like flood plains, coastal zones and areas vulnerable to earthquake and volcano eruptions. This would also include providing early warning signals, shelter and adequate relief facilities, promoting group action by citizen groups, to play a constructive role in emergency.

(ix) To conserve the earth's vitality and diversity, taking actions needed to:

- Preventing pollution.
- Restoring and maintaining the integrity of the Earth's eco-systems.
- Conserving biological diversity.
- Ensuring that renewable resources are used sustainably.
- Devising appropriate Earth Care Policies to meet the above objectives in regard to preventing pollution. These will include economic and regulatory measures for reducing emissions of sulphur dioxide, nitrogen oxides, carbon monoxide and hydero carbons.
- Preparing for climate change, induced by the addition of greenhouse gases and International Environmental Consensus involving principles and rules of collaboration among soverign states arrived at by common consensus.
- Restoring, increasing and maintaining the integrity of the Earth's Natural and Modified Systems, through an integrated approach to land and water management, using the drainage basin as the unit of management.
- Adopting scientific land use planning in all natural eco-systems—forests, wetlands, scrublands and grasslands as well as any high-quality farm land.
- Restoring and rehabilitating degraded eco-systems through appropriate scientific measures.
- Assigning the highest priority to more sustainable agriculture practices in collaboration with the local people, in areas suffering from serious losses of soil erosion, desertification and misuse of land.
- Halting net-deforestation and protecting large areas of old-growth forest areas and the governments establishing explicit long-term goals in national forest policy.
- Involving local communities in forest policy formulation as well as in their execution and management to ensure sustainability of yields (Joint Action).
- Conserving biological diversity by seriously pursuing a combination of measures to safeguard species of genetic stocks, through the establishment and maintenance of

protected areas, and adoption of wider strategies that combine economic activities and conservation, with people's participation. For maintaining species and genetic resources, *in situ* and *ex situ* conservation should be practiced.

- Keeping population within the Earth's carrying capacity is not merely *just* for survival, but for ensuring sustainable improvement in the quality of life of the people. Earth Care Policies must pay due importance to both these aspects and also to the issues of *population growth* and *sustainable consumption* in an integrated way.
- Distortionary subsidies and prices must be removed.
- The prices of resources should match with their real costs.
- To inculcate the sanctity of Mother Earth and as a "*Thanks-Giving*" method, "*Earth-Day*" should be celebrated all over the country in Schools, Colleges, Universities, Offices and business establishments. This will motivate "*Greener Movements*" and ultimately pave the way to "*Earth Democracy*", with sustainable development and sustainable life styles, as its core policies.
- If an "*Earth Democracy Movement*" becomes a reality, an Earth Party can emerge as the governing party, with Earth Care Policies as a primary item in its manifesto and uniting and bringing all the multiple political parties in the country towards a common and noble cause.
- An '*Earth Party*' will be a change agent and a potent force for bringing about a cumulative set of '*green actions*' for sustainable resource consumption, introducing sustainable life styles and ensuring sustainable development. This development may sound utopian, but not unattainable in the New Millennium, as our Earth is fast racing towards its death.

Our effort should be to build a '*Sustainable and Heavenly Earth*', which can last through eternity. I conclude this paper with a vision and dream expressed in a Poem titled, "*Heavenly-Earth—a Prayer*", which is an adapted version from Rabindranath Tagore's *Gitanjali*.

HEAVENLY EARTH—A PRAYER

Where the mind is without fear and the head is held high;
Where knowledge is free;
Where the world has not been broken up into fragments by narrow domestic walls;
Where the Earth retains its greenery, sanctity and natural beauty;
Where birds in freedom sing and chirp without fear;
Where all living species and life support systems co-exist without threat and extermination;
Where man's greed does not overtake his need;
Where the Earth's diversity and vitality are preserved and protected;
Where man lives in perfect harmony with the bounties of nature, within the carrying capacity of the Earth;
Where man listens to the cry of an ailing Earth "*Avani-Dhwani*" and acts with
veneration and respect for natural laws;
Where man uses Vasundhara's resources with restraint and wisdom;
Where the Earth's precious resources and fragile ecosystems are managed with
prudence and care for a sustainable future;
Where humans desist from causing air and water pollution and increasing
economic toxicity;
Where tireless striving stretches its arms towards conservation, preservation and Scientific management of natural resources;
Where governments, U.N. entities, the private sector and the civil society march in unison, towards the goal of sustainable development;
Where man learns to produce and consume wisely and sustainably, and holds the Earth as a '*Treasure in Trust*' for the future generations;
Where a benign Science and Technology is truly harnessed for promoting sustainable development and a liveable planet;
Where human leadership is able to rise above the narrow domestic walls of caste,
religion, community and exploitation of Man by Man;
Where mega-cities, metropolises, slums and shanty towns yield place to a

Micropolitan-centred sustainable human settlement pattern of
development;
Where Nations share the bounties of Mother Earth in just and
equitable proportions in the global system;
Where women's potentials are harnessed for environmental
governance, through emancipation and participation in national
and International ecosystems management, and control of
environmental degradation;
Where the unsustainable use of our finite natural resources is
effectively controlled to reduce their depletion;
Where "*water wars*" are averted by ensuring Water and Food
security through a steady supply of water for safe drinking and
productive irrigation;
Where equity and justice in the use and control of water are
ensured through people-centred Watershed development approach
and inter-linking of river basins;
Where "Village Swaraj", with the just devolution of powers and
finances, manages all local affairs and the Civil society takes over
the management, and control of local projects with active public
participation;
Where the voices of the poor and the disadvantaged groups are
listened to with justice and freedom;
Where the clear stream of reason has not lost its way into the
dreary desert sand of dead habit;
Where the mind is led forward by THEE into ever-widening
thought and action towards sustainable development and
sustainable lifestyles;
Into that heavenly earth, my Father, let our ecosystems awake.

(*Adapted from Rabindranath Tagore's Gitanjali*)

NOTES

1. For instance, the average yield per hectare of rice in China is about twice that in India, while in the case of wheat it is 1-1/2 times higher.
2. Ravindra Babu, A. "Combating Land Degradation and Droughts" in **Rio-Johannesburg and Beyond,** LEAD INDIA, Orient Longman Pvt. Ltd. 2002, Hyderabad.
3. Poor and Sayer 1987; *The Management of Tropical Moist Forest Land: Ecological Guidelines*, IUCN Gland, Switzerland.

4

Integrating Traditional and Scientific Knowledge: An Approach to Earthcare

REKHA SINGHAL and NIDHI SHARMA

Introduction

Rapid degradation of forests, depletion of water table, pollution of water and degeneration of the problems are faced by the humanity other natural resources, attempts are being made towards their sustainable conservation and management. Since the early 1980s, several approaches towards sustainable natural resource management, which includes active participation of local communities, have been attempted in many countries. In this regard, the role of Indigenous/ Traditional Knowledge (TK) in sustainable natural resource management has received increased interest and attention during last few years, not only by the researchers but also from policy makers, project designers, and implementing agencies. The subject has outgrown from the image of local boundary and is being discussed and debated at global level through the documents of international organizations (viz. World Bank initiative on 'IK' for development and UNESCO's database of 'best practices'), networks, and workshops. Studies on knowledge and management practices, which traditional people have used for ages, and which have been continuously adapted to environmental changes, reveal a wealth of information on forest development and water conservation.

Indigenous Knowledge

Warren (1993) considers IK synonymous to 'traditional' and 'local' knowledge. According to him, IK refers to the knowledge of indigenous people as well as any other defined community. More recently, Grenier (1998) in her guide for researchers, working with IK, defined IK as, "The unique, traditional, local knowledge existing within and developed around the specific conditions of women and men indigenous to a particular geographic area".

A more elaborated definition of IK considers 'IK' as the sum total of knowledge and skills which people in a particular geographic area possess, and which enable them to get the most out of their natural environment. Most of this knowledge and skills have been passed down from earlier generations, but individual men and women in each generation adapt and add to this body of knowledge in a constant adjustment to the changing circumstances and environmental conditions. They in turn pass on the body of knowledge intact to the next generation, in an effort to provide them with survival strategies" (Marrewijk, 1998).

Formal/Scientific Knowledge

Scientific knowledge is taught in universities, both in developing and developed countries. It is based on scientific research, the result of experiments and extensive testing under different conditions. This knowledge is usually introduced onto the communities by external organisations such as national forest departments, but it is not related to the communities' world-view or other cultural aspects (Leeuwen, 2000).

Comparison between Traditional and Scientific Knowledge

In view of achieving the goal of sustainable natural resource management, it is increasingly recognised that exclusive management by a single approach does not ensure achievement of objective. There is growing consensus that the traditional knowledge is constantly evolving. Its enrichment will depend upon whether it can interact with new types of information and can be integrated with scientific knowledge to address the emerging issues. Table 4.1 presents a comparison between traditional and scientific knowledge.

Table 4.1: Comparison between Traditional and Scientific Knowledge (After Singhal R., 2000)

Parameters	*Traditional Knowledge*	*Scientific Knowledge*
Context	Local area specific	Broad based, general
Framework	Integrated	Specialisation
Roots	Society	Laboratory controlled condition.
Time frame	Location specific sequence of information	Widespread information system
Approach	Process begins with community/people	Process begins from expert/external sources
Strategy	Locally appropriate action leading to increase in local capacity	Promotes dissemination and transfer of knowledge derived from expert/external organisations
Observations	Mostly qualitative/subjective	Mostly quantitative and subjective
Attitude	Respects learner's ideas and knowledge	Very little consideration of learners participation
Values	Does not claim to have answer to problems rather helps the process of the change to move	Makes knowledge claims and attempts to impose them on community
Records	Human mind	Print media
Extension	Mostly informal and oral	Mostly formal and written
Problem	Defined by the internal and owner of the knowledge	Generally defined by the external parties
Outcome	Strengthens belief and sense of local community	Weakens local decision making and sense of detachment

It is generally acknowledged that participation of stakeholder communities in management of their resources is essential for the credibility and sustainability of interventions and at the sametime contributes to the empowerment of communities. In this respect, emphasis on incorporation and usage of traditional knowledge in interventions, thus enhancing active participation of indigenous communities and allowing them to present their views and ideas, contributes to the success and sustainability of the interventions, to empowerment of indigenous people and to the overall knowledge base regarding natural resource management.

Integration of Traditional and Scientific Knowledge through Adaptive Learning

The sustainable management of natural resources involves making and implementing decisions about the use and maintenance of these resources. The decision making process involves integration and utilisation of several kinds of knowledge. These different kinds of knowledge come from both scientific and traditional sources and can be combined through a process of adaptive learning, drawing on the principles of adaptive management. Adaptive management is essentially a process of 'learning by doing' as described by Bos (1974). The principles of adaptive management are recognized as the key ingredient of adaptive learning. Adaptive management "deals with the unpredictable interactions between people and ecosystems as they evolve together" (Berkes and Folke, 1998).

On the basis of principles of adaptive management (Holling, 1998), a model of adaptive learning is proposed.

Integration of Knowledge Through Adaptive Learning for Earth Care

Adaptive learning depends upon appropriate integration of knowledge and its verification (or testing of knowledge). Based on the assumption that the natural resource management is governed by the knowledge base, which is continuously verified, adapted and accepted by the people, a two dimensional model of adaptive learning is presented in Figure 4.1.

VERIFICATION	Knowledge: Scientific	Knowledge: Traditional
ON OTHERS	1 Sectoral & Technical Based	2
ON SELF	3	Holistic & Societal Based 4

Figure 4.1: A Model of Adaptive Learning (After Singhal R., 2000)

Thus, at the heart of adaptive learning theory (Figure 4.1) there are basically two critical dimensions, i.e. knowledge and verification.

Knowledge is a critical dimension in determining adaptive learning. It includes information, values, believes and attitudes. Knowledge differs at least in two important ways: Scientific (Technical) and Traditional (Indigenous). Verification of knowledge deals with action or action sequences, which provides adaptation, acceptance and feedback. Verification of knowledge can be done either on self or on others.

Natural resource management may be characterized as involving interaction between resource and the users. Figure 4.2 illustrates that the scientific knowledge is generally technical-based, sectoral, specialized and verified on others. Therefore, for adaptation, need assessment, and acceptance, feedback from the user is a necessary condition. On the other hand, traditional knowledge moves through the trial and error process on self, resulting in practical, dynamic, need-based, adaptable, integrated and societal-based knowledge during the phase of knowledge development itself.

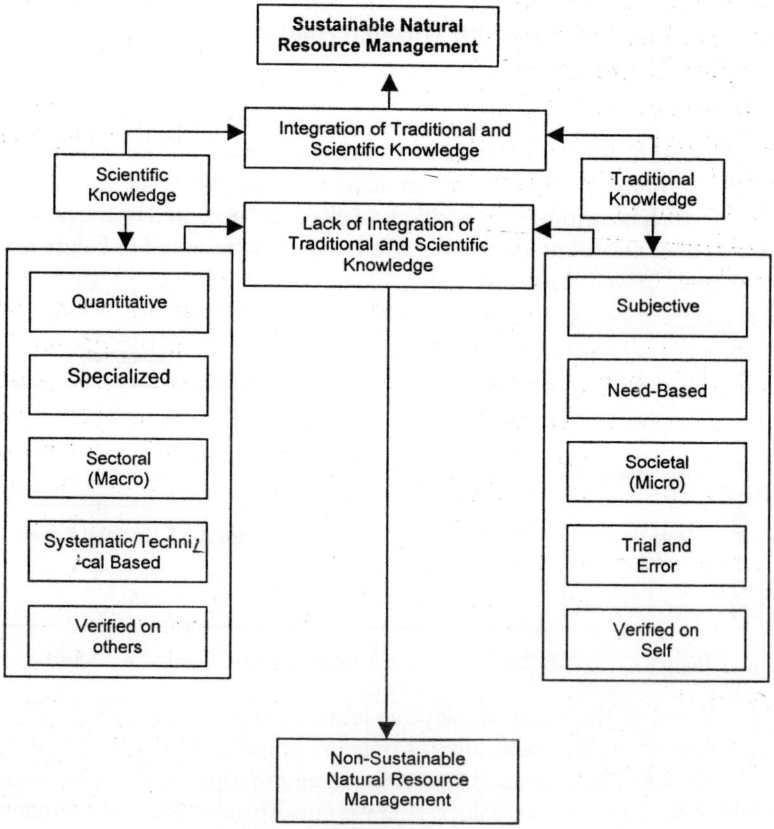

Figure 4.2: Conceptual Framework Integrating Traditional and Scientific Knowledge.

It is important to note that this natural resource management related knowledge that communities have learned by doing, is closely associated with their culture. This reveals that it is not only a valuable contribution to the formal ecological knowledge, but also the up-to-date information on the present condition of the environment, its changes in the past and the actors that include these changes, serves as framework for sustainable natural resource management.

A look at Figure 4.2 makes it evident that to achieve sustainable natural resource management integration of traditional and scientific knowledge is essential. This integration will help to reduce the limitations and will facilitate maximization of benefits of each knowledge system.

Local Forest Management in India

In the middle of 1970s, in India, the expansion of informal local forest protection movement eventually triggered the issuing of a national resolution in June 1990. The resolution provided the authority to local communities to participate in management of state forest and specifies their rights over forest produce. This movement is popularly known as Joint Forest Management.

Traditional Forest Management: Selected Case from Himalayas

Local people (Bhotias) in the Central Himalayan Region of India traditionally make conscientious use of natural resources for their daily needs and subsistence livelihood. They directly depend on nearby forests, vegetation for their daily needs such as fodder, firewood, and timber for household needs, medicinal plants, fruits, etc. Their dependence on natural resources has been increasing with the increase in population and due to immigrants from planes. However, land being limited resource, increased pressure on land started effecting the ecological balance and conservation of natural resources. It was found that religious believes markedly influence the manner, in which a human society perceives and utilizes its natural resource. It was noted that various religious resource, conservation points were established by the local community to conserve a maximum number of species of varied utility. This innovative practice is serving as an effective means of conservation and is in total harmony with the environment, which have persisted through generation (Sharma, Rikhari, and Palni, 1999).

The extensive examples of forest management by local people are also found in eastern and northeastern parts of India. In these hilly upland and tribal regions, communities have shown remarkable capacity for managing their land and natural resources. It was found that several kinds of organizations were formed at the village level to take up the management of forest. For Example, in Budhikhanmari, Orissa, where more than 55 villagers were jointly carrying out forest protection on the basis of principles of equity for which, they have received several awards' was unique in this aspect. However, it was found that most of the committees were struggling under financial constraints (Kant *et al.*, 1991; Jonsson and Rai, 1994).

Local Water Management in India

The depleting level of water table and excessive exploitation of underground water and massive deforestation has forced the policy planners to evolve a strategic management system, encompassing different models of development. Rajiv Gandhi Watershed Mission is one such programme, which was started in various states with the aim of involving local people and professionals in management and conservation of water resource.

Traditional Water Management: Selected Case from Madhya Pradesh

Karondi Milli Watershed Programme was started in the Katni district of Madhya Pradesh with the main objective of conservation and optimum utilization of five "j" of environment, i.e. "jal", "jamin", "jungle", "jan", "jeevan". In order to implement the programme, a survey and analysis of the area was conducted and emphasis was laid on adoption of traditional method of water conservation and people's participation. Different works were undertaken by the project implementing agencies which included soil and land management, contour trenching, check dams, crop management, rural energy management, fisheries, horticulture health, human resource development etc. Integrating technology, social organization and administrative arrangements (decentralization of decision making) were crucial to the programme success. In devising and adopting sustainable means of development strategy participatory approach was adopted. In order to involve local people, sufficient ground was prepared through organizing training camps, workshops and mass campaign. During the course of the programme, it was realized that participation of watershed committee members in discussions with village people helped in of some common understanding of the problems and their potential solutions. Involvement of people was also facilitated by the fact that locally available technologies were put to use rather than imposing external technology. At the sametime local seasonal and crop calendar was taken into consideration. Community consolidation and community action in a participatory way was prepared for capacity building.

The success of watershed development programme is fully dependent on efficiency of management and effectiveness of people's participation because it gives an impetus to the working process and makes the programme a self-sustained one. Similarly, use of commonly available resources and identifying the key areas where the available technology can be put to use provides a useful link in bridging between the gap between the haves and have-nots (Tripathy and Sahu 2002).

A Model of Pro-Environmental Action

Pro-environmental action depends on an appropriate integration of affective and cognitive dimension. When people are aware of the environmental issues, understand the importance of the natural resource, and are emotionally attached to them, their behaviour will be directed towards pro-environmental action and will lead to sustainability.

COGNITION \ Affective	No	Yes
No	1 Passivity	2 Sensitivity
Yes	3 Activity	4 Sustainability

Figure 4.3: A Model of Pro-Environmental Action

At the heart of the pro-environmental action theory there are basically two psychological concepts, awareness and affection.

- *Cognition*—Cognition is a critical dimension in determining pro-environmental action. It can be defined as a mental concept, including awareness, motivation, information, knowledge and locus of control.

- *Affection*—Affection is related to the emotional aspect of an individual. It includes person's feelings and values, towards general and specific issues and has a strong impact on a person's behaviour/actions.

Figure 4.3 presents the basic model of pro-environmental action. As shown in the figure, the four parts indicate—passivity, activity, sensitivity and sustainability. Whereby each cell represents four major possible combinations and illustrates the outcome in terms of:

- People and individuals are not aware of the environmental issues and there is no attachment for the natural resource. This situation leads to passivity.
- People are emotionally attached to the natural resource, but there is no information/awareness about their conservation/protection. This situation leads to sensitivity whereby people do not have idea about actionable steps.
- People are only involved in activity for some motive/need but have absolutely no attachment to the natural resource. This situation leads to activity, which stops as soon as the need or motive is fulfilled.
- People are attached to the natural resource as well as they have information and awareness regarding the importance and the ways to protect their natural resource. This in turn will facilitate sustainability of their pro-environmental actions.

Conclusion

The incorporation and usage of traditional knowledge in interventions enhance active participation of communities, hence contributing to the success and continuity of interventions, to empowerment of indigenous people and to the broadening of the overall knowledge base.

The overwhelming literature on indigenous communities and their management related knowledge indicates that there is a need to work out methodologies, that can be applied to support indigenous people. This article has put forward a framework and classification that may provide an insight into various opportunities, to facilitate learning from indigenous people and integration of Traditional and Scientific Knowledge, through the principles of adaptive learning. Application of both indigenous and formal knowledge, complimenting to each other can have a two-fold advantage in interventions: Communities are more likely to accept and natural resource management that can benefit from best aspects of both. In this respect, incorporation of traditional knowledge can serve to enhance communication between formal professionals and communities and contribute to communities' empowerment and self-esteem.

REFERENCES

Berks, F., & Folke C., (1998). *Linking Social and Ecological Systems for Resilience and Sustainability.* In Berks, F. & Folke, C. (eds.) *Linking Social and Ecological Systems: Management Practices and Social Mechanisms for Building Resilience.* Great Britain: Cambridge University.

Bos, A. H., (1974). *Oordeelsvorming in Groepen: Willens en Wetens, Wikken en Wegen; Polariteit en Ritme als Sleutel tot Ontwikkeling Van Sociale Organismen.* (Ph.D. Dissertation: Wageningen Agricultural University, The Netherlands). WAU Dissertation Abstract No. 586.

Grenier, L. (1998). *Working with Indigenous Knowledge: A Guide for Researchers.* Ottawa: IRDC.

Jonsson, S. & A. Rai. (1994). *Forests, People and Protection: Case Studies of Voluntary Forest Protection by Communities in Orissa.* ISO/Swedforest, New Delhi.

Holling, C. S. (1998), *Science, Sustainability and Resources Management.* In Berkes, F. & Folke, C. (eds.) *Linking Social and Ecological Systems: Management Practices and Social Mechanisms for Building Resilience.* Cambridge University. Great Britain.

Kant, S., N. M. Singh & K. K. Singh. (1991). *Community Based Forest Management System: Case Studies from Orissa.* ISO/Swedforest Indian Institute of Forest Management and SIDA, Bhopal.

Marrewijk, A. V. (1998). *Indigenous Knowledge: The Proof is in the Eating of the Pudding.* Indigenous Knowledge and Development Monitor, (6), 12-13.

Sharma, S., H. C. Rikhari, & L. M. S. Palni, (1999). *Conservation of Natural Resources Through Religion: A Case Study from Central Himalaya.* Society and Natural Resources, (12), 599-612.

Singhal, R., (2000). *A Model for Integrating Indigenous and Scientific Forest*

Management: Potentials and Limitations for Adaptive Learning. Forestry, Forest Users and Research, ETFRN, Wageningen, The Netherlands.

Singhal, R. (2000). *Learning from Resource Users—A Paradigm shift in Tropical Rorestry?* Paper presented in Workshop April 2000, Vienna, Austria.

Tripathy, O. P. & Sahu, Bimla. (2002). *Milli Watershed Development, The Karondi Experience*, Kurukshetra, New Delhi.

Warren, D. M. (1993). *Background to the International Symposium on Indigenous Knowledge and Sustainable Development.* Paper presented in September 1992 in Silang, Philippines.

5

Sustainable Relationship Between Forest and Agriculture: A Case of the Gir Wildlife Sanctuary

R. L. SHIYANI, H. R. PANDYA and I. U. DHRUJ

Introduction

Forests constitute an important natural resource base in India. The forests in India are said to cover 76.5 million hectares (mha), which constitute 23 per cent of the total Indian geographical area of 328 million hectares, whereas the National Forest Policy resolution aims at having minimum of one-third of the total area of the country under forest or tree cover. It was also clear that, of the 23 per cent of the total geographical area classified as forests, forest cover varied in density. According to the Forest Survey of India (FSI, 1998), 48 per cent area notified as forests had a crown density of more than 40 per cent, 34 per cent between 10 and 40 per cent, and the rest 18 per cent had less than 10 per cent or no tree cover at all. Forests are also not spread evenly in India but are concentrated more in the poorer regions of low agricultural productivity and poor soils with heavy tribal concentration.

Forests in Gujarat

Gujarat with a total landmass of 196024 sq. km has been endowed with plenitude of bio-diversity and natural ecosystems. The area under forests is 18868.28 sq. km, i.e. 9.63 per cent which is much less than the national average of 23.4 per cent of geographical area. The actual forest cover in the state is only 12044 sq. km (6.1 %) and rest of the forest area is devoid of vegetation. The land use pattern of the state is depicted in the pie chart given below.

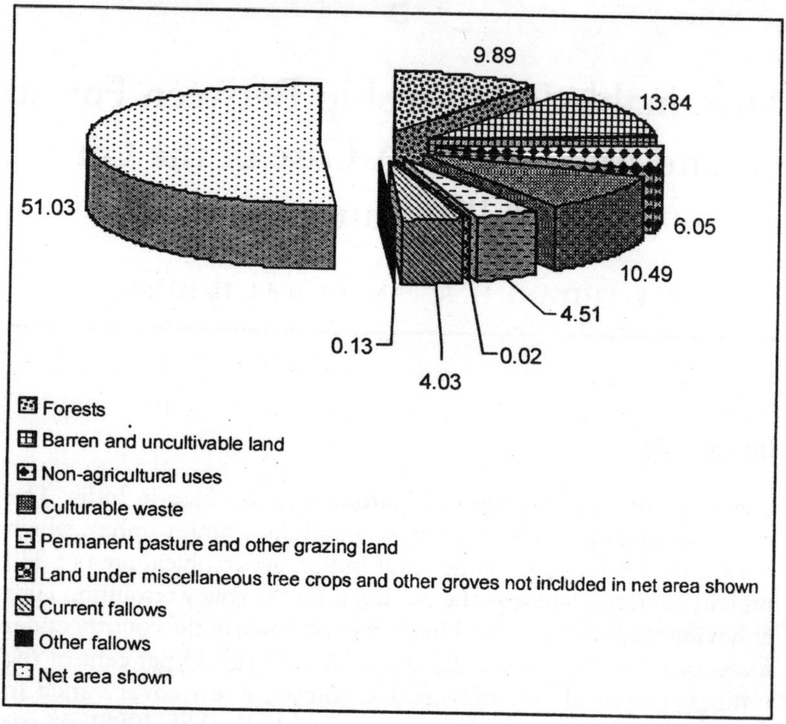

Source: Verma (1999).

The Gir forest in the Kathiawar Peninsula of Gujarat has been famous for the Asiatic lion, *Panthera leo persica*. The Government of Gujarat has declared an area of 1412.12 sq. km as Protected Area (PA) which comprises 258.72 sq. km as National Park and 1153.40 sq. km as Sanctuary. Besides this, 470.50 sq. km forest area constitutes buffer zone as reserve, protected and unclassed forest. Thus, a total area of 1882.62 sq. km forms Gir Forest. Out of the total area of 1882.62 sq. km 1502.72 sq. km falls in Junagadh district and 379.90 sq. km in Amreli district (Meena *et al.*, 1998). The Gir lies between 20°40′ N and 21°50′ N latitude and 70°50′ E and 71°50′ E longitudes.

Despite very high realisation of the importance of forests and wildlife, the economic value of forests have not caught the public attention to the extent it deserves. The goods and services provided by forests are not tangible and it is really difficult to assign economic value to intangible benefits. Moreover, the forest resources are also open and hence access and utlisation by the local people go unrecorded. Thus, the valuation of the environment is anthropocentric and the associated economic costs and benefits lead to a conflict between diverging interests of conserving the natural resources and the need to exploit them for developmental activities. Since the economic analysis of forest resources is difficult, there exists a dearth of literature on quantification of environmental and economic benefits of conserving forests. This study, therefore, aims to quantify the environmental and economic benefits of conserving Gir eco-system which would be of great help to policy makers, planners, foresters and researchers.

Analytical Framework

Estimation of benefits either environmental or economic is a complex procedure for any eco-system because the forest has a variety of products and uses, the market value of some of which is difficult to evaluate. Valuing these benefits is an important task for quantifying economy-environment inter-relationships. These benefits can be quantified and valued using various approaches.

In the environmental economics parlance, total economic value (TEV) concept is used to cover all direct and indirect values; actual use value, option values and existence and bequest values. The same was used in the present study. The efforts were made to quantify as far as possible all the identified benefits employing TEV technique. The total economic value can be expressed as:

$$TEV = DUV + IUV + OV + EV$$

Where,

DUV = Direct use value
IUV = Indirect use value
OV = Option value and
EV = Existence value

The flow chart indicating this method and major benefits of a protected area is depicted in Figure 5.1. Using this method, the valuation of some of the use value and non-use value is discussed below:

Use Value

This value arises from the direct or indirect consumption of goods and services of a Gir eco-system. If the consumers are willing to pay for some of the benefits for future use, an option value is generated and that can also be considered in the use values. The use values were, therefore, classified into three groups, viz., direct, indirect and option values.

Direct Use Values

These values were generated from the use of the resources and services rendered directly by the Gir eco-system. The collection of forest products such as fuel wood, fodder, fruits, etc. by local communities is an example of consumptive direct use value. On the other hand, the tourism, research and education etc. are the examples of direct non-consumptive use value. Viewing lions in the wild is the best example of direct non-consumptive use values. Direct use values computed for the study included: (i) Benefits to agricultural sector, (ii) Grass production, (iii) Tourism, (iv) Nature education, (v) Employment generation in mango orchard and sugar industries, (vi) Fuel wood, and (vii) Timber.

Indirect Use Values

These comprise mainly the environmental functions of the Gir eco-system. Recharging of ground water, raising crop productivity, the eco-system services such as fixing of nitrogen, carbon store, gene pool, etc. are all indirect use values. In the present study, the indirect use values considered were: (i) Environmental (Quantifiable and non-quantifiable), (ii) Research and education, (iii) Water resources and (iv) Recreation.

Option Values

It is associated with the benefits received by retaining the option of using a resource in the future by protecting or protecting it today, when its future demand or supply is uncertain. The bio-diversity is an example of optional value.

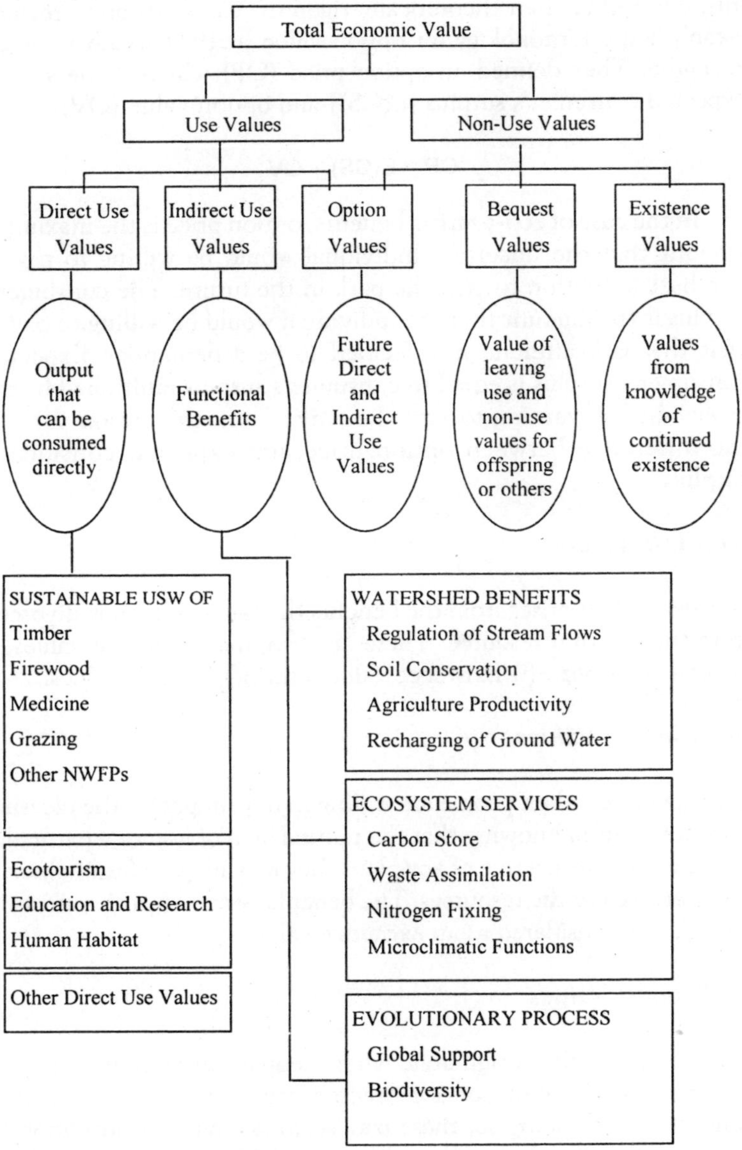

Figure 5.1: Total Economic Value of Protected Areas

Weisbrod introduced the concept of option value in 1964. But Weisbrod was not precise in his definition and his arguments were intuitive rather than theoretically rigorous. Cicchetti and Freeman established a terminology for option value in 1977 which is widely accepted. They defined an option price (OP) which is the sum of expected consumer's surplus [E(CS)] and option value (OV).

$$OP = E(CS) + OV$$

In the case of eco-tourism benefits, option price is the maximum amount that the uncertain individual would be willing to pay to purchase an option to visit the park in the future. The consumer's surplus is the amount that the individual would be willing to pay to visit the park after he has decided to be a demander. Expected consumer's surplus is equal to consumer's surplus multiplied by the probability of wanting to visit the park. Thus, the option value is the difference between option price and expected consumer's surplus.

Non-Use Values

Non-use values arises from the benefits that are completely divorced from the use of a resource. There are two major non-use values of natural areas, viz., (i) Existence values and (ii) Bequest values.

(i) Existence Values

This generates when people are willing to pay simply for the pleasure they derive from knowing that the natural area of lion or a particular specie exists irrespective of any plans they may have to hunt, observe or otherwise use the resources. The benefits derived by the tourists in Gir PA was considered as an existence value.

(ii) Bequest Values

The bequest value originates, when people are willing to pay to conserve the benefits of a protected area for the use of future generations. By doing so, these people do not have an intention to 'use' the benefits during their life span, but are bequesting those benefits for the future generations. The details of various research studies and nature education camps conducted in Gir PA were considered as bequest value.

Therefore TEV = UV + NUV
= DUV + IUV + OV + BV + EV

Where,

DUV = Direct use value
IUV = Indirect use value
OV = Optional value
BV = Bequest value
EV = Existence value

The goods and services provided by a PA can be divided into two groups for valuation. One is marketed goods and the other is non-marketed goods. In the case of marketed goods, market price may not always be an adequate economic value due to market failure. In such a context, shadow prices are usually taken as the surrogate market price. Among non-marketed goods, certain goods and services may have a close substitute in the market which could help the valuation process. However, a large number of goods and services provided by PA have no market at all. The methodology to estimate the economic value of non-market goods is described in the form of flow chart (Figure 5.1).

Costs of Gir Eco-System

The total costs of conserving Gir eco-system are divided into two broad heads: direct cost and opportunity cost. The direct costs of the Gir eco-system are the budget outlays for the establishment and other expenditure on various programmes including compensation for human and animal causalities caused by carnivorous. The data were obtained from the various financial reports of the Forest department. The opportunity costs are the benefits foregone by choosing the next best alternative.

Tourism

Gir forest has gained reputation as an important wildlife tourist spot in the country. Gir receives three kinds of visitors, viz., (i) those visiting to see the lion, (ii) travellers passing through Gir *via* state roadways and (iii) the pilgrims. There are 26 temple complexes where more than 1.5 lakh pilgrims visit annually. Of these, three temples may be considered as the major ones: Tulsishyam, Kankai and Banej, which together account for most of the pilgrims traffic.

The amenities for tourists like accommodation, guides and interpretive facilities etc. are made available by Forest Department. Apart from these, a hotel of Taj group, Maneland Tourist Complex and Umang are situated in Sasan Gir. The Dharmshalas (shelter houses) run by the temples are also available in three temples located in Gir PA.

The income generated by the Forest Department, through the fees collected from the tourists visited the Gir PA is given in Table 5.1. Its compound growth rates are presented inTable 5.2. The periodwise data of the tourists visited indicated that the annual average number of tourists visited the Gir PA was more or less same upto 1989-90 (Table 5.1). But it rapidly increased in the current decade. The average annual number of tourists visited during the period 1990-91 to 1999-2000 was about 2.7 times to those visited in earlier two periods. As a result, the income from tourists visited increased manifolds in the current decade. The annual average income generated from the tourists during the current decade was to the tune of Rs. 232.40 million. Tiwari (2000) also observed that the rising number of tourists in Gir has resulted in more revenue and promoted the local economy. The people visiting Gir PA are also potential asset in terms of spreading the message of conservation (Nath, 1997; Singh and Srivastava, 1997).

Table 5.1: Income from Tourists Visited Gir PA

Sl. No.	*Year*	*Total tourists visited (No.)*	*Average tourists visited per annum (No.)*	*Total income (Rs.)*	*Average Income (Rs.)*
1.	1977-78 to 1979-80	77485	25828	1132447	377482
2.	1980-81 to 1989-90	261041	26104	4386213	438621
3.	1990-91 to 1999-2000	700206	70021	23240016	2324002
4.	1977-78 to 1999-2000	1038732	45162	28758676	1250377

An attempt was also made to work out the willingness to pay (WTP) for the tourists' visit to prime places like Sasan and Safari Park, Devaliya including religious places, viz., Tulshishyam, Kankai and Banej. Thirty visitors were interviewed for the amount they wanted to sacrifice for their visits. It was found that the visitors were not interested to pay any additional amount for visiting religious places but they were keen to visit prime places even by showing their willingness to pay very high amount. It was estimated from the number of tourists visited for the period from 1990-91 to 1999-2000, that the total annual additional amount of Rs. 0.54 million would be sacrificed by all the tourists. This amount was added in direct use value while computing the total economic value of the Gir PA.

Table 5.2: Compound Growth Rates of Number of Tourists Visited the Gir PA and Revenue Generated

Sl. No.	Tourists	Parti-culars	Unit	1977-78 to 1989-90	1990-91 to 1999-2000	1977-78 to 1999-2000
1.	Indian	Mean	Number	25662	68109	44117
		CGR	Per cent/annum	1.79	13.67*	7.70*
2.	Foreigners	Mean	Number	378	1911	1045
		CGR	Per cent/annum	6.95	18.96*	13.87*
3.	Total	Mean	Number	26040	70020	45162
		CGR	Per cent/annum	1.90	13.81*	7.81*
4.	Revenue from tourists	Mean	Rupees	424512.2	2324002	1250377
		CGR	Per cent/annum	1.61	28.56*	12.30*

* Indicates significant at 5 per cent level of significance.

CGR = Compound Growth Rate.

The annual production figure for timber and firewood are given in Table 5.3. along with an estimated figures drawn for Gir PA. The average wood biomass production in entire Gujarat is assumed to be equally distributed. The total timber and firewood quantity is 145700 and 1311300 cmt, respectively for entire Gujarat forest area of 18868.28 sq. km. The total value for the same is Rs. 1295.46 million. At present, the area of Gir PA is about 1412 sq. km. By applying simple arithmetic calculation, the estimated total timber and firewood quantity is 10903.40 and 98130.60 cmt, respectively. The total estimated value is Rs. 96.94 million which is eight per cent of entire state value for wood biomass. Meena *et al.* (1998) reported that nearly 15000 mt fuel wood was collected only by the villagers and *maldharis* from Gir ecological unit. Hence, the present estimation for Gir PA gets support from this observation.

Table 5.3: Estimation of Timber and Firewood Production from Gir PA

Item	Quantity (cmt/year)		Rate (Rs.)	Value (Million Rs.)	
	Gujarat	Gir PA		Gujarat	Gir PA
Timber	145700	10903.40	4000/cmt	582.80	43.61
Firewood	1311300	98130.60	750/mt	712.66	53.33
Total	1457000	109034.00	—	1295.46	96.94

N.B: 1.38 cmt = 1mt.
Source: Verma (1999).

Estimation of Environmental Benefits from Gir PA

The estimation of environmental benefits of Gir PA was worked out as per Forest (Conservation) Act (1980). The Rules and Guidelines were amended on October 25, 1992. As per this Act, environmental value of one hectare of fully stocked forest (Density 1.0) would be taken as Rs. 12.674 million to accrue over a period of 50 years, i.e., per square kilometre the value of benefit would be Rs. 1267.40 million. These environmental benefits include soil erosion prevention, effect on hydrological cycle, wildlife habitat, microclimate, upsets of ecological balance etc. For estimation of forest density in Gir PA, the data of forest cover under different vegetation type were utilized as reported by Indian Space Research Organization, Ahmedabad (ISRO, 1999). Since exact density was not given, the density for various kinds of vegetation was assumed for the purpose. The area in sq. km in respective vegetation was calculated on the basis of total area of Gir PA, i.e. 1412.12 sq. km. The estimated benefit in million Rs./sq. km was worked out as per the formula given below:

$$\text{Annual Environmental Benefit (Rs. million/sq. km)} = \frac{1267.4 \times \text{Forestd density}}{50}$$

Data presented in Table 5.4 revealed that the maximum area of 445.66 sq. km. of Gir PA was occupied by dry deciduous with full density vegetation type, followed by the dry deciduous open forest. The density assumed for both these vegetation is 0.50 and 0.20, respectively. The highest density of 0.8 was assumed for riverine dense forest (113.82 sq. km). Maximum total annual benefit of Rs. 5646.51 million at 1992 price was obtained from dry deciduous full density vegetation type. It was followed by riverine dense forest (Rs. 2308.27 million). The total estimated annual benefit from Gir PA accrued was Rs. 10575.71 million. The exact estimation of environmental benefit in monetary term was not available for Gir PA. Verma (1999) has also estimated the environmental services rendered by Gujarat forests following almost similar methodology with the same base year. He valued Rs. 139439 million for 18868.28 sq. km forest of entire Gujarat state. On the basis of this figure, the value for Gir PA (1412.12 sq. kms.) would be Rs. 10435.75 million. The estimation made in this paper is roughly equal to the estimation made by Verma (1999). This justifies the methodology followed in this study for estimating the environmental benefits from Gir PA.

Table 5.4: Estimation of Environmental Annual Benefits* from Gir PA

S. No.	*Vegetation Type****	*Assumed Forest Density*	*Per cent Area****	*Area in sq. km*	*Benefit (Rs. in million/sq. km)***	*Total Benefit (Rs. in million)*	*Value at 1999-00 price (Rs. in million)*
1.	Riverine dense forest	0.80	08.06	113.82	20.28	2308.27	3346.99
2.	Forest with moist species	0.60	02.31	032.62	15.21	496.15	719.42
3.	Dry deciduous with full density	0.50	31.56	445.66	12.67	5646.51	8187.44
4.	Dry deciduous open forest	0.20	26.45	373.50	05.07	1893.64	2745.78
5.	Dry deciduous mixed degraded	0.10	06.47	091.36	02.53	231.14	335.15
	Total					10575.71	15334.78

* Include soil erosion, effect on hydrological cycle, wild life habitats, microclimatic uses.

** Value of one hectare (density 1.0) Rs. 126.74 lakhs to accrue over a period of 50 years based on 1992.

*** *Source*: ISRO (1999).

Note: The value for the year 1999-2000 was calculated taking the base year of 1993-94 = 100. The source is RBI bulletin, August 2000, pp. S-793.

Expenditure on Gir PA

Keeping in view the significance of maintaining Gir PA, the Forest Department spends huge amount under various heads of plan and non-plan expenditure. These data were, therefore, collected and reported in this section.

It is interesting to note that the number of visitors from other countries (foreigners) increased manifolds in the current decade compared to earlier periods. The results of compound growth rates (CGR) revealed that the annual rate of compound growth for Indian as well as foreign visitors was very high and significant in the current period compared to earlier periods. It is also pertinent to note that the annual rate of increase in foreign visitors was higher than the Indian visitors in all the periods (Table 5.2). This leads to infer that Gir PA attracted foreigners to a greater extent compared to Indian tourists, which clearly shows the keen interests of foreigners in Gir forest.

Table 5.5: Average Non-plan Expenditure for Gir PA (Average of 1995-2000)

Sl. No.	*Budget head*	*Annual average in Rs.*	*Per cent to total*	*C.V. %*
1.	Divisional Offices	19372365	32.43	54.23
2.	Jodi guards	466122	0.78	19.40
3.	Communications	609252	1.02	46.37
4.	Construction of grass godown	676511	1.13	33.87
5.	Building	903814	1.51	48.04
6.	Improvement and Extension of forest	614904	1.03	27.41
7.	Preliminary Expenditure on exploitation of forest produce	3392	0.01	48.83
8.	Exploitation of forest produce by government agency	2196278	3.68	12.23
9.	Individual beneficiary scheme	43861	0.07	48.38
10.	Scheme for nature conserving including conservation of wildlife	2564783	4.29	18.48
11.	Management of sanctuary and National park	3851006	6.45	14.44
12.	Project for development of Gir and Barda lion sanctuary	1211581	2.03	21.12
13.	Development of Gir and Barda lion sanctuary	4901375	8.21	7.88
14.	Development of Gir and Barda lion sanctuary	455878	0.76	93.80
15.	Scheme for wildlife education interpretation and training	182118	0.30	23.59
16.	Establishment of wildlife safari park in detached area	174746	0.29	25.31
17.	Eco-development in and around N.P. and Sanctuary	119241	0.20	223.60
18.	Eco-development project for Gir N.P.	21262590	35.60	88.84
19.	Demarcation and Survey	120895	0.20	223.60
	Total	59730712	100.00	50.76

Table 5.5 presents the average of five years non-plan expenditure on Gir PA. The total average annual non-plan expenditure was Rs. 59.73 million. The break-up of total expenditure indicated that the expenditure on eco-development project had maximum share (35.60 %), followed by the expenses of Divisional Offices (32.43%), development of Gir Sanctuary (8.21%), management of sanctuary and National Park (6.45%), etc. On an average, about 51 per cent variability in the total non-plan expenditure was found in the last five years. Maximum variability was found in the expenditure on eco-development project and demarcation, and survey.

In addition to regular expenditure (non-plan), the long term investment, i.e. plan expenditure also plays crucial role in any developmental activity. The data on plan expenditure for the period from 1995-2000 were obtained from the annual reports of Forest Department. These data were averaged out to nullify abnormal fluctuation. Thus, the data on average plan expenditure for Gir PA are presented in Table 5.6. On an average, the total annual plan expenditure for Gir PA was to the tune of Rs. 27.16 million. The break-up of total plan expenditure showed that the integrated forestry development project had a lion's share (62.93%) in the total plan expenditure, followed by management of sanctuary and National Park (10.03 %), development of Gir Sanctuary (6.99 %), development of communication (4.10 %), etc. Relatively less variability was noticed in the total annual plan expenditure (31.26 %). However, wide range of variability among the different cost components was observed. It varied between 27.87 per cent in case of expenditure on development of Gir sanctuary and 223.61 per cent in respect of expenses on scheme for nature and environment training to officer and staff.

Table 5.7 presents last five years data on plan and non-plan expenditure for Gir PA. It can be observed from the data that there was rapid increase in both the types of expenditures upto 1997-98. During the years 1998-99 and 1999-2000, the plan expenditure has increased marginally as compared to the year 1997-98. On the other hand, the non-plan expenditure though increased continuously, the rate of increase was marginal after 1997-98. Similar trend was noticed for total expenditure for Gir PA. On an average, the total annual expenditure for maintaining Gir PA was Rs. 86.89 million.

Opportunity Cost of Gir PA

The opportunity cost is the value which one has to sacrifice in order to maximize the returns or for any alternate purpose. The total area under Gir forest is 141212 hectares and the net sown area was estimated to be 103819.06, i.e. 73.52 per cent of the total forest area (Table 5.8). On the basis of net profit earned per hectare by the sample households beyond 25 km from Gir PA, the total opportunity cost of Gir forest land was worked out to be Rs. 1181.47 million. This amount is added to the total cost of maintaining Gir, because the forest land could have been utilized for agricultural purpose to earn the income if there had not been forest.

Table 5.6: Average Plan Expenditure for Gir PA (Average of 1995 to 2000)

Sl. No.	*Sub-budget head*	*Annual average in Rs.*	*Per cent to total*	*C.V.%*
1.	Plantation of medicinal plants	1272	0.00	137.30
2.	Individual beneficiary scheme	34290	0.13	161.12
3.	Demarcation and Survey	567141	2.09	93.73
4.	Forest protection	277443	1.02	95.89
5.	Scheme for nature and environment training to officer and staff	51513	0.19	223.61
6.	Management of sanctuary and N.P.	2724993	10.03	33.15
7.	Development of Gir and Barda Lion Sanctuary	1898199	6.99	27.87
8.	Wildlife education interpretation. and training	338233	1.25	89.54
9.	Scheme for marine national park	58396	0.22	32.14
10.	Scheme for minimizing and printing depreciation by wildlife to human life and property	292722	1.08	187.61
11.	Development of wildlife outside forest area	279933	1.03	144.81
12.	C.S.S. Plantation of minor forest produce	135266	0.50	90.20
13.	Development of communication	1114766	4.10	52.70
14.	Construction of building	119627	0.44	223.60
15.	Soil and Moisture conservation and afforestation in area	1015344	3.74	137.31
16.	Plantation of coastal boarder	234830	0.86	102.65
17.	Integrated forestry development project	17091630	62.93	49.73
18.	Compensatory afforestation	31408	0.12	90.88
19.	Development of forest settlement	390086	1.44	67.75
20.	Development of forest settlers	191467	0.70	78.45
21.	Ayurved research survey and herbs garden	114668	0.42	95.21
22.	Grass collection and watchmen charges	197617	0.73	223.60
	Total	27160844	100.00	31.26

Table 5.7: Average Plan and Non-Plan Expenditure on Gir PA

(*In. Rs.*)

Sl. No.	*Year*	*Plan*	*Non-plan*	*Total*
1.	1995-96	19088604	12628147	31716751
2.	1996-97	37084140	32114662	69198802
3.	1997-98	71563089	34145166	105708255
4.	1998-99	80508379	27987257	108495636
5.	1999-2000	90409337	28928987	119338324
6.	Average	59730710	27160844	86891554

Table 5.8: Opportunity Cost of Gir PA Forest Land

Sl. No.	*Particulars*	*Unit*	*Total*
1.	Total Forest area	ha.	141212.00
2.	Net sown area	ha.	103819.06 (73.52 %)
3.	Profitability	Rs./ha	11380.10
4.	Total opportunity cost of Gir PA	Rs. million	1181.47

Note: Figure in bracket is the per cent of net sown area to total forest area.

Total Economic Benefits from Gir PA

As discussed in the methodology chapter, Total Economic Value (TEV) is relatively better tool for quantifying the benefits from Gir PA. Using this approach, direct use value, indirect use value, other benefits and costs were worked out. The details are given in Table 5.9.

It is observed from the table that the grazing, eco-tourism fees, timber, firewood, benefits from Nature Education Camps (NEC), value of agricultural production, sugar industries and mango marketing are classified as direct use value. Thus, the annual direct use value generated from these was Rs. 7039.46 million. Similarly, environmental benefits were considered as indirect use value. The total indirect use value per annum was estimated to be Rs. 15334.78. The total annual benefits from direct use value and indirect use value was to the tune of Rs. 22374.24 million. Various costs involved in maintaining Gir PA were also worked out. It includes all the management costs and the opportunity cost of forest land. The total annual such costs were Rs. 1268.36 million. Total economic value of Gir PA was computed by deducting the total costs from the total benefits. Thus, the annual total economic value of entire Gir PA was as high as Rs. 21105.88 million. This huge amount of benefits clearly suggests that the maintenance of Gir PA for the present and also for the future generations should be accorded high priority by all the concerns.

Table 5.9: Total Annual Economic Benefits from Gir PA

(Rs. in million)

Sl. No.	Particulars	Item	Income	Expenditure
1.	*Direct use value*	Grass	680.20 (9.66)	—
		Tourists' fees	2.32 (0.03)	—
		Willingness to pay (Tourist)	0.54 (0.01)	—
		Timber	43.61 (0.62)	—
		Firewood	53.33 (0.76)	—
		Benefit from NEC	1.19 (0.02)	—
		Agricultural production	6020.05 (85.52)	—
		Sugar industries	166.98 (2.37)	
		Marketing of mango	71.24 (1.01)	—
		Sub-total	7039.46 (100.00)	—
2.	*Indirect use value*	Environmental benefit	15334.78 (100.00)	
3.	*Total benefit*		22374.24	—
4.	*Management cost*		—	86.89 (6.85)
5.	*Opportunity cost of forest land*		—	1181.47 (93.15)
	Sub-total			1268.36 (100.00)
6.	*Total economic annual benefits (Rs. in million)*		21105.88	

Note : Figures in parentheses are per cent of the total.

REFERENCES

Forest (Conservation) Act, (1980). *Rules and Guidelines* (As amended on October 25, 1992). Government of India. Ministry of Environment and Forest, New Delhi.

Forest Survey of India (FSI). (1998). *State of India's Forests*, 1997, Dehradun.

ISRO, (1999). Baseline mapping of Gir PA, Gujarat using remote sensing and GIS techniques. Gir PA, Sasan, Final report. Forest and Environment Division (FED) FLPG/RESA. Space Application Centre (ISRO), Ahmedabad.

Meena, R. L., Kamboj, R. D. and Singh, M. (1998). "*The* Gir", Gir Welfare Fund, Wildlife Division, Sasan Gir.

Nath, T. (1997). Evaluate the potential available infrastructure and tourism management practices in Gir conservation unit (GCU) and suggest a framework for tourism management. Management term papers presentation, *Gir Conservation Unit* (GCU), 21st March, 1997.

Singh, M. and Srivastav, A. (1997). *Tourism plan for Gir Forest Development*, Government of Gujarat, 1997.

Tiwari, R. (2000). Impact of tourism on wildlife management and the local economy in Gir. XXI post-graduate diploma course in wildlife management, Wildlife Institute of India, Dehra Dun.

Verma, A. K. (1999). Contribution of Gujarat forests to the state economy. Booklet published by Publicity Liaison Division, Gandhinagar, p. 14.

6

Population and Land Use: Concerns for Sustainable Development

DEWARAM A. NAGDEVE

Introduction

As the Twenty-first century begins, the rapid increase of human numbers combines with desperate poverty and rising levels of consumption are depleting natural resources on which the livelihood of present and future generations depends. One of the major causes of environmental degradation in India could be attributed to rapid growth of population, which is adversely affecting the natural resources and environment. The growing population and the environmental deterioration face the challenge of sustained development without environmental damage. To achieve an expansion of national output, it is essential to combine natural resources, human resources and capital. The existence or the absence of favourable natural resources can facilitate or retard the process of economic development. The Cairo Conference in 1994 have helped in evolving a holistic viewpoint where population, development and the environment are seen as inextricably interlinked, each influencing the others, and where human development is the ultimate goal in tackling these factors. Spiral linkages amongst excessive population growth, extremes of poverty and the effects of increasing population pressure on the natural resources and environment have been the main areas of concern for demographers, ecologists, environmentalists, development planners and policy makers over the recent past.

The sustainable development implies long term sustainability in all economic sectors including industry, energy, agriculture, forestry and consumption. Environmentalists and economists increasingly agree that efforts to protect the environment and to achieve better living standards can be closely linked and are mutually reinforcing. Slowing the increase in population, especially in the face of rising per capita demand for natural resources, can take pressure off the environment and buy time to improve living standards on a sustainable basis (Green, C. P., 1992).

Though the relationship is complex, population size and growth tend to expand and accelerate these human impacts on the natural resources and environment. All these in turn lead to an increase in the pollution levels. However, environmental pollution not only leads to deteriorating environmental conditions but also have adverse effects on the sustainable development and health of people. Further, the population will increase to such an extent in future that it will cause overall scarcity for resources. India is having 16.87 per cent of the world's population on 2.4 per cent of its land area has great deal of pressure on its all natural resources including land. The increase of population has been tending towards alarming situation. The world's population was estimated to be 6.056 billion in mid-2000 and projected 7.936 billion in the year 2025 and 9.322 billion in the year 2050. Contribution of India alone to this population was estimated to be 1009 millions in mid-2000 which has been projected 1352 millions in 2025 and 1572 million in the year 2050 (World Population, 2000). The projected population indicates that India will be a first most populous country in the world and China will be second in 2050. If the world population continues to multiply, the impact on environment could be devastating. The number of population rise will increase to such an extent in future that it will cause overall scarcity for resources. The deficiency of natural resources will create environmental hazards in the world. India, with about 16 per cent of the world's population on only 2.4 per cent of its land area has great deal of pressure on its all natural resources including land. Decades of economic expansion and population growth have degraded its land, air and water.

India is one of the most degraded environment countries in the world and it is paying heavy health and economic price for it. According to a World Bank sponsored study, estimated environmental damage in the year 1992 amounted to about US $ 10 billion or Rs. 34,000 crores, which is 4.5 per cent of GDP. Urban air pollution costs India US $ 1.3 billion a year. Water degradation leads to health costs amounting to US $ 5.7 million every year, nearly 60 per cent of the total environmental cost. Soil erosion affects 83 to 163 million hectares of land every year. Beside, land degradation leads to productivity loss equal to US $ 2.4 billion or 4 to 6.3 per cent of the agricultural productivity every year (UNDP, 1998).

The present paper seeks to examine some of the linkages between population change, increasing urbanization and its influence on natural resources, and the environment.

Population Growth in India

India is the second most populous country in the world after China. Recently, the population of India has crossed the one billion marks. At the time of independence, the country's population was 342 million. The number has multiplied three-fold in around five decades. The selected demographic characteristics of the population of India are presented in Table 6.1. The total population size of India had grown from 361 million in 1951 to around 846 million in 1991. According to the 2001 census, the population of India is 1027 million. The population of India increased by more than 2.3 times during the period of 1951-91. It is estimated that the country's population will increase to 1.26 billion by the year 2016. The rural population of India has increased more than double from 298.7 to 628.7 million during 1951-91, whereas the urban population has grown 3.5 fold from 62.4 million in 1951 to 217.6 million in 1991. The decadal growth rates of the population are irregular, as it increased from 13.31 per cent in 1951 to 24.8 per cent in 1971. It marginally declined to 24.7 per cent in 1981 and 23.8 per cent in 1991. The rural decadal growth rate of population varies from 8.79 per cent in 1951 to 20.01 per cent in 1991, whereas the urban decadal growth rate of population varies from 41.43 per cent in 1951 to 36.47 per cent in 1991. There are various reasons for this variation in the trend of population growth rate in various censuses. The increase in population has been due to the improvement in health conditions and control of diseases. The density of population has gone up from 117 in 1951 to 267 persons in 1991 and it always shows a increasing trend over the census years in persons per square kilometre. Several push and pull factors are presumed to be operative towards distress out migration from rural to urban areas. This might be due to the declining resource availability per capita and shrinking economic opportunities in rural areas, and better economic opportunities, health and educational facilities etc. in urban areas, providing opportunities for higher level of human capital development could be the underlying factors for rural outmigration.

Table 6.1: Selected Demographic Characteristics of the Population of India, 1951-2001

Year	Population in Million			Decadal Growth Rate of Population			Density of Population	Ratio of Population to 1951 Population
	Total	Rural	Urban	Total	Rural	Urban		
1951	361.1	298.7	62.4	13.31	8.79	41.43	117	100
1961	439.2	360.3	78.9	21.51	20.49	26.41	142	121.63
1971	548.2	439.1	109.1	24.8	21.86	38.23	173	151.81
1981	683.3	523.9	159.4	24.66	19.32	46.14	216	189.23
1991	846.3	628.7	217.6	23.85	20.01	36.47	267	234.37

Source: Census of India, Provisional Population Totals, 2001.

Vital Rates in India

The growth of population depends upon fertility, mortality and migration. The process of accelerated population growth in India till 1970s were observed to witness a marginal deceleration during 1980s as the decadal population growth was 23.9 per cent compared to 24.9 per cent in 1970s. Further declines in fertility have been witnessed during 1990s. The estimated birth, death, natural growth, infant mortality and total fertility rates in India is presented in Table 6.2. It is revealed from table that birth rate in India has declined from 33.9 per thousand populations in 1981 to 29.5 per thousand populations in 1991 and it further declined to 26.4 per thousand populations in 1998. At the same time the crude death rate has also declined from 12.5 per thousand populations in 1981 to 9.8 per thousand populations in 1991 and it further declined to 9 per thousand populations in 1998. Thus, the annual natural growth rate of India's population has declined from 1.97 per cent in 1991 to 1.74 per cent in 1998. The total fertility rate has also declined from 4.52 children per women in 1981 to 3.32 children per women in 1997. Infant Mortality Rate (IMR) per thousand in India has steadily declined from 110 in 1981 to 72 in 1991 but the present IMR (72) is still very high.

Table 6.2: Estimated Birth, Death, Natural Growth, Infant Mortality and Total Fertility Rate in India, 1981-1998

Year	CBR	CDR	NGR	TFR	IMR
1981	33.9	12.5	2.14	4.52	110
1986	32.6	11.1	2.15	4.15	96
1991	29.5	9.8	1.97	3.64	80
1992	29.2	10.1	1.91	3.6	79
1993	28.7	9.3	1.94	3.54	74
1994	28.6	9.2	1.94	3.57	73
1995	28.3	9	1.93	3.54	74
1996	27.5	9	1.85	3.4	72
1997	27.2	8.9	1.83	3.32	71
1998	26.4	9	1.74	NA	72

Source: Sample Registration System Bulletins, Registrar General, India.

Trends in Poverty and its Environmental Effects in India

Most of India's poor live in rural areas and are engaged in agriculture. India's poverty reduction through the anti-poverty and employment generation programmes along with overall economic growth planning efforts have helped to reduce the poverty ratio in the country. The trends in poverty in India are depicted in Table 6.3. The people below poverty line declined from 55 per cent in 1973 to 36 per cent in 1993-94. The absolute number of poor have, however, remained constant at around 320 million over the years. During the same period the fraction of population below poverty line dropped from 56.4 per cent to 37.3 per cent in rural areas and from 49 per cent to 32.4 per cent in urban areas. Over the period 1987-88 to 1993-94, urban poverty declined but rural poverty has remained broadly unchanged. The only viable long term strategy for poverty eradication is (a) increasing incomes of those engaged in agriculture and (b) absorbing an increasing proportion of the labour force into industry and services.

Poverty is said to be both cause and effect of environment degradation. The poor people, who rely on natural resources more than the rich, deplete natural resources faster as they have no real prospects of gaining access to other types of resources. Poorer people, who cannot meet their subsistence needs through purchase, are forced to use common property resources such as forests for food and fuel, pastures for fodder, and ponds and rivers for water. Clean drinking water facility through taps is available to only 35 per cent of urban households and 18 per cent of rural households in India. Other residents use unsafe water sources like wells, ponds and rivers. Population pressure driven overexploitation of the surface and underground water resources by the poor has resulted into contamination and exhaustion of the water resources. Urban population is also using rivers to dispose of untreated sewage and industrial effluent. The result is that health of those dependents on untreated water resources is increasing by at risk. In the absence of capital resources, the poor are directly dependent on natural resources. Moreover, degraded environment can accelerate the process of impoverishment, again because the poor depend directly on natural assets. Acceleration in poverty alleviation is imperative to break this link between poverty and the environment. The poverty and rapid population growth are found to coexist and thus seems to reinforcing each other. It also contributes to environmental degradation through over exploitation of natural resources like land, air and water. The deterioration of natural resources and unsafe living conditions affects the environment and health of the poor people.

Table 6.3: Number and Percentage of Population Below Poverty Line in India: 1973-1993

Year	Rural		Urban		Combined	
	Number in Million	Poverty Ratio	Number in Million	Poverty Ratio	Number in Million	Poverty Ratio
1973-74	261	56.4	60	49.0	321	54.9
1977-78	264	53.1	65	45.2	329	51.3
1983	252	45.7	71	40.8	323	44.5
1987-88	232	39.1	75	38.2	307	38.9
1993-94	244	37.3	76	32.4	320	36.0

Source: Planning Commission Estimates.

The creation of employment opportunities is essential in agricultural areas with high poverty, unemployment and landlessness. Poverty, by reducing incentives for long term resource management, can deplete the natural resources. Environment degradation, in turn, lessens the labour productivity of poor and the productivity of natural resources they manage. Poor people to meet their needs overexploit the natural resources. Poverty also affects the demographic characteristics of the population and hinders the transition to slower population growth.

Land Utilization Patterns in India

A change in land utilization pattern implies an increase or decrease in the proportion of area under different land uses at a point in two or more time periods. Table 6.4 describes the land utilization pattern in India from 1951 to 1991. It shows variations in land use and a narrow range of fluctuations in the proportion of net sown area to total land in the country since 1951 to 1991. Out of total geographical area of 329 million hectares, only 305 million hectares is the reporting area (the rest being unadministered for various reasons). The land for non-agricultural uses (housing, industry and others) is increased from 9.36 million hectares in 1951 to 22.35 million hectares in 1991. About 19 million hectares are snow bound and remote leaving only 263 million-hectare for agriculture, forestry, pasture and other biomass production. The area under cultivation had increased by about 30 per cent until 1981 and thereafter depicts marginal decline. The net sown area increased from 119 million hectares in 1950-51 to 140 million hectares in 1970-71 mostly through reclamation of old fallow and culturable wastelands and diversion of groves. The net area sown has increased only marginally from 140 million hectares in 1970-71 to 142 million hectares in 1995-96, indicating that the private efforts have peaked and the intervention of the government is required for further land reclamation. In fact this area has marginally declined by 0.74 million hectares in 1995-96 than that of 1994-95.

The extent of agricultural intensification and extensification characterized by increase in cropping and irrigation intensity and higher use of chemical fertilizers, pesticides and insecticides. The process of agricultural extensification and intensification is leading to land degradation, overexploitation of underground water resources, increased use of chemical fertilizers leading to eutrophication and water pollution. Agricultural intensification because of increasing cropping intensity, irrigation intensity and excessive use of chemical fertilizers resulting into water logging, salinization and alkalinization of croplands and eutrophication of water bodies and ill health of oceans and thus reductions in biodiversity.

Table 6.4: Land Use Patterns in India, 1951-1996

Classification	*1950-51*	*1960-61*	*1970-71*	*1980-81*	*1990-91*	*1995-96F*
I. Geographical Area	328.7	328.7	328.7	328.7	328.73	328.73
II. Reporting Area for land utilization statistics (1 to 5)	284.32	298.46	303.76	304.15	304.86	304.89
1. Forests	40.48	54.05	63.91	67.47	67.8	68.83
2. Not available for cultivation (a+b)	47.52	50.75	44.64	39.62	40.48	41.38
(a) Non-Agricultural Uses	9.36	14.84	16.48	19.66	21.09	22.35
(b) Barren and Unculturable land	38.16	35.91	28.16	19.96	19.39	19.03
3. Other Uncultivated Land (excluding fallow land) (a+b+c)	49.45	37.64	35.06	32.31	30.22	28.64
(a) Permanent pasture and other grazing land	6.68	13.97	13.26	11.97	11.4	11.05
(b) Land under miscellaneous tree crops and grooves not included in net area sown	19.83	4.46	4.3	3.6	3.82	3.49
(c) Culturable wasteland	22.94	19.21	17.5	16.74	15	14.1
4. Fallow Land (a+b)	28.12	22.82	19.88	24.75	23.36	23.82
(a) Fallow land other than current fallows	17.44	11.18	8.76	9.92	9.66	10.01
(b) Current Fallows	10.68	11.64	11.12	14.83	13.7	13.81
5. Net area sown	118.75	133.2	140.27	140	143	142.22
6. Gross cropped area	131.89	152.77	165.79	172.63	185.74	186.56
7. Area sown more than once	13.14	19.57	25.52	32.63	42.74	44.35
8. Cropping intensity*	110.1	114.7	118.2	123.3	129.9	131.2
III. Net irrigated area	20.85	24.66	31.1	38.72	47.78	53.51
IV. Gross irrigated area	22.56	27.98	38.19	49.78	62.47	71.51

P: Provisional, *: Cropping Intensity is obtained by gross cropped area by net area sown.

Source: Department of Agriculture and Cooperation, Ministry of Agriculture Forest Resources.

Forest Resources

Forests are an important natural resource of India. They play an important role in providing raw materials to industries and generating income and employment. Forests also play an important role in enhancing the quality of environment by influencing the ecological balance and life support system (checking soil erosion, maintaining soil fertility, conserving water, regulating water cycles and floods, balancing carbon dioxide and oxygen content in atmosphere etc. They have moderate influence against floods and thus they protect the soil erosion.

India has a forest cover of 76.52 million square km of recorded forest area, while only 63.34 million square km can be classified as actual forest cover. These accounts for 23.28 per cent of total geographic area against 33 per cent recommended by National Forest Policy of 1988. Per capita availability of forests in India is much lower than the world average. In the year 1997, as compared to 1993, the total forest cover has decreased by 6710 sq. km. The states, which have shown significant decline in the forest covers, are Andhra Pradesh and Madhya Pradesh. Whereas the states of Gujarat, Maharashtra, Rajasthan and West Bengal have shown an increase in forest cover (Table 6.5).

The total area under forests was 64.08 million hectares in 1981-83 which was about 19.5 per cent of the total geographical area and it declined to 63.34 million hectares or 19.3 per cent in 1993-95, as against the National Forest Policy 1988 stipulation of a target of 33 per cent. Even within this recorded area, only 36.7 million hectare, or only 11.2 per cent of country's total land area, comprises dense forest with a crown density of more than 40 per cent, thus reflecting a qualitative decline of forests in the country. The total forest area diverted for non-forestry purposes between 1950 and 1980 was 4.5 million hectares, i.e. at an annual rate of 0.15 million hectare. To regulate unabated diversion of forestland for non-forestry purposes, Forest (Conservation) Act, 1980 was enacted. It has resulted in reduction of diversion of forest area for non-forestry purposes considerably and the present rate of diversion is 16,000 hectare annually (Economic Survey of India, 1998-99). The devastation caused to the economy by deforestation is responsible for greater frequency and intensity of floods, soil erosion, heavy siltation of dams built at enormous expense and changes in climatic conditions. Continuing deforestation, therefore, has brought us face to face with a major ecological and socio-economic crisis.

Table 6.5: Forest Cover Estimates for India, 1993-1997

State	Area in Million Square Km		Absolute Change	Per cent Change
	1997	1993		
Andhra Pradesh	43.29	47.26	–3.97	–8.39
Bihar	26.52	26.59	–0.06	–0.24
Gujarat	12.58	12.04	0.53	4.43
Haryana	0.60	0.51	0.09	17.74
Himachal Pradesh	12.52	12.50	0.02	0.15
Karnataka	32.40	32.34	0.06	0.19
Kerala	10.33	10.34	0.00	–0.02
Madhya Pradesh	131.20	135.40	–4.20	–3.10
Maharashtra	46.14	43.86	2.28	5.21
Orissa	46.94	47.15	–0.20	–0.43
Punjab	1.39	1.34	0.04	3.28
Rajasthan	13.35	13.10	0.25	1.94
Tamil Nadu	17.06	17.73	–0.66	–3.73
Uttar Pradesh	33.99	33.96	0.03	0.10
West Bengal	8.35	8.19	0.16	1.99
ALL INDIA	633.40	640.11	–6.71	–1.05

Source: Forest Survey of India. The State of Forest Report, 1997.

Land/Soil Degradation

Direct impacts of agricultural development on the environment arise from farming activities, which contribute to soil erosion, land salination and loss of nutrients. The spread of green revolution has been accompanied by over exploitation of land and water resources and use of fertilizers and pesticides have increased many folds. Shifting cultivation has also been an important cause of land degradation. Leaching from extensive use of pesticides and fertilizers is an important source of contamination of water bodies. Intensive agriculture and irrigation contribute to land degradation particularly salination, alkalization and water logging. It is evident that most of the land in the country is degrading, thus affecting the productive resource base of the economy. Out of the total geographical area of 328.7 million hectares, 175 million hectares are considered to be land-degraded area (Table 6.6). Water and wind erosion is the major contributor of

141.3 million hectares to soil erosion, with other factors like water logging 8.5 million hectares, alkali soil 3.6 million hectares, acid soil 4.5 million hectares, saline soil including coastal sandy areas 5.5 million hectares adding to the situ degradation. While soil erosion by rain and river in hill areas causes landslides and floods, deforestation, overgrazing, traditional agricultural practices, mining and incorrect siting of development projects in forest areas have resulted in opening up of these areas to heavy soil erosion. Ravines and gullies reported 4 million hectares, area subject to shifting cultivation reported 4.9 million hectares and riverine and torrents erosion due to floods and eutrophication due to agricultural run off reported 2.7 million hectares. The increasing intensification and extensification also results in salination, alkalization and water logging in irrigated areas of the country. For achieving and maintaining food security, sustainable forestry agricultural and rural developments controlling of land/soil erosion is very much necessary.

Table 6.6: Soil Erosion and Land Degradation

(Million Hectares)

1. Total Geographical Area	328.7
2. Area Subject to Water and Wind Erosion	141.3
(Area Degraded through Special Problems)	
3. Water Logged Area	8.5
4. Alkali Soil	3.6
5. Acid Soil	4.5
6. Saline Soil including Coastal Sandy areas	5.5
7. Ravines and Gullies	4.0
8. Area subject to Shifting Cultivation	4.9
9. Riverine and Torrents	2.7
Total 3 to 9	33.7

Source: Economic Survey of India, 1998-99, Government of India.

Most part of the applied pesticides and fertilizers, irrespective of crop, applicator or the formulation used, ultimately finds its way into the soil. Before pesticides are completely inactivated, they may adversely affect the functioning of non-target microbes and other forms of life inhabiting the soil. They may also be taken up by the plants or get translocated in the aquatic system by leaching or run-off, thus contaminating the plankton, fish, invertebrate and other forms of life using their water. Pesticide residues in food items have been a matter of considerable concern. Even small quantities of these residues ingested daily along with food can build up high levels in the body fat. The long term effects of these residues in the human body include carcinogenecity, reduced life span and fertility, increased cholesterol, high infant mortality and varied metabolic and genetic disorders (c.f. Compendium of Environment Statistics, 1999).

Traces of pesticides and fertilizers from the fields are washed into the nearest water bodies at the onset on monsoons or whenever there are heavy showers that add to water pollution. Consumers are affected by agricultural concomitants such as pesticides and fertilizers that run-off from fields into rivers. Polluting a river is dangerous because generally, rivers are the primary source of drinking water for towns and cities downstream of the point of pollution beside other sources. The New Agricultural Policy seeks to sensitize the farming community with the environmental concerns. The policy lays that concerted efforts will be made to pool, distil and evaluate traditional practices, knowledge and wisdom and to harness them for sustainable agricultural growth (Downloaded from Internet, New Agricultural Policy, http\\www.nic.in).

The increasing river water pollution is the biggest threat to public health. The diseases commonly caused due to polluted water are cholera, diarrhoea, hepatitis, typhoid, amoebic and bacillary, dysentery, guineaworm, whereas scabies, leprosy, trachoma and conjunctivitis are some of the diseases associated with water scarcity. All these could be attributed to the rapidly increasing population and lack of water resources. Inadequate access to safe drinking water and sanitation facilities leads to higher infant mortality and intestinal diseases. More than one million children died due to diarrhoea and other gastrointestinal disorders in 1990. In addition, around 90 lakh cases of acute diarrhoea diseases have been reported in India. Uttar Pradesh reporting the highest number of cases (Central Bureau of Health Investigation, 1996). It is estimated that 73 million workdays are lost every year due to water related diseases. The cost of treating them and the loss in production amount to Rs. 600 crores a year (Citizen's Report, 1982).

The World Health Organization (WHO) has projected as to what proportion of water related diseases could be done away if proper water supply and sanitation services are provided. It can be observed from Table 6.7 that diseases such as cholera, typhoid, leptospirosis, scabies, dracunculiasis can be reduced by 80 to 100 per cent by improvements in the supply of water and sanitation services. Reductions in other water borne diseases are also possible up to an extent of 40 to 70 per cent.

Table 6.7: Potential Reductions in Morbidity for Different Diseases as a Result of Improvements in Water Supply and Sanitation

Diseases	*Potential Reduction in Morbidity (%)*
Cholera, Typhoid, Leptospirosis, Scabies, Dracunculiasis	80-100
Trachoma, Conjuctivitis, Yaws, Schistomiasis	60-70
Tularaemia, Paratyphoid, Bacillary, Dysentery, Gastroenteritis, Louse borne diseases, diarrhoeal diseases, Ascariasis, Skin diseases	40-50

Source: Intersectoral Action for Health, Geneva, WHO, 1986.

Summary and Conclusions

Population growth process in India got accelerated until 1970s and depicted slight deceleration during eighties. The outcomes of high population growth rates are increasing population density, pressure on natural resources and increasing number of people below poverty line. Indian population has multiplied three-fold and the share of rural population has declined whereas the share of urban population has been increased in around five decades. Density has increased more than doubles over the period. The concerted efforts are required to bring down fertility below replacement level. The death rate has been declining faster as compared to birth rates but infant mortality is still high.

The land use pattern in India does not depict substantial changes over period characterized by trends in croplands, pasturelands and forestlands during 1950-51 to 1995-96. The percentage of people below poverty line declined but the absolute number of people below poverty line increased from 1973 to 1993-94. Over the period 1987-88 to 1993-94, urban poverty declined but rural poverty has remained broadly unchanged. The only viable long term strategy for poverty eradication is (a) increasing incomes of those engaged in agriculture and (b) absorbing an increasing proportion of the labour force into industry and services. The creation of employment opportunities is essential in agricultural areas with high poverty, unemployment and landlessness. Poverty also affects the demographic characteristics of the population and hinders the transition to slower population growth.

The study reveals that rapid population growth has led to the overexploitation of natural resources. The deforestation has led to the shrinking of forest cover, which eventually affects human health. The devastation caused by deforestation is responsible for greater frequency and intensity of floods, droughts, soil erosion, heavy siltation of dams built at enormous expense and changes in climatic conditions. However, environmental pollution not only leads to deteriorating environmental conditions but also have adverse effects on the sustainable development and health of people. The considerable amount of both ground water and surface water contamination due to chemical fertilizers and insecticides in the country leads to various water borne diseases. Intensive agriculture and irrigation contribute to land degradation particularly salination, alkalization and water logging. It is evident that most of the land in the country is degrading, thus affecting the productive resource base of the economy. The increasing intensification and extensification also results in salination, alkalization and water logging in irrigated areas of the country. For protecting natural resources and environment, sustainable forestry, rural developments, controlling of water contamination due to chemical fertilizers, pesticides, land/soil erosion, is very much necessary.

From the various effects of environmental degradation on human beings, discussed in this paper, it appears that if human beings wants to exist on earth, there is now high time to give top priority to protect natural resources and environment. There is a need control pollution of all types for a healthy living. There is a need to control population growth below replacement level in the country. Special efforts should be made for informing and educating the people and local leaders about the adverse effects of large population through specially designed IEC (Information, Education and Communication) activities. In order to increase green cover and to preserve the existing forests; afforestation and social forestry programmes should be implemented at the local level. There is a need for preventive and curative measures to control water pollution due to chemical fertilizers, pesticides and other wastes. Wastewater treatment plants should be established in accordance with the need of time and its usage should be encouraged. The heavy penalty should be imposed on industries disposing off the wastes into the river. Moreover, the landfills are to be properly managed to prevent ground water contamination. More emphasis should be laid on compulsory environmental education at the school level in order to make people aware of the environment protection. The environment protection should not be a responsibility of government alone but local people and leaders should be encouraged to make dedicated efforts to eradicate the environmental problems.

To sum up, there is an urgent need to control population and conserve and protect natural resources and protect the environment for sustainable development and for the betterment of present and future generations.

REFERENCES

Carter, Brandon and Kirsten Honmann, (1991-92). "Valuing Environmental Costs in India: The Economy Wide Impact of Environment Degradation", World Bank, mimeo.

Central Bureau of Health Intelligence, (1995 & 1996). Health Information of India, Ministry of Health and Family Welfare, Government of India, New Delhi.

Central Statistical Organisation, (1998 & 1999). *Compendium of Environment Statistics*, Department of Statistics, Ministry of Planning and Programme Implementation, Government of India.

Central Statistical Organisation, (1998). Proceedings of the National Workshop on Environment Statistics, 1998, 12-13 January, Goa, Department of Statistics, Ministry of Planning and Programme Implementation, Government of India.

Central Statistical Organisation, (2000). Proceedings of the National Workshop on Environment Statistics, 2000, 6-7 April, Hyderabad, Ministry of Statistics and Programme Implementation, Government of India.

Government of India, (1999). *Economic Survey: 1998-99*, Ministry of Finance, Economic Division.

Green, C. P. (1992). *The Environment and Population Growth: Decade for Action*, Population Reports, Series M, No. 10, Vol. XX, No. 2, Baltimore, Johns Hopkins School of Public Health, Population Information Program.

Gulati, S. C., (2000). *Population-Development-Environmental Linkages in India*, Paper presented in Millenninm Conference on Population, Development and Environment Nexus organised by IASP, PFI and UNFPA at New Delhi, February 14-16, 2000.

Hinrechsen, Don and Robey Briant (2001). *Population and the Environment: The Global Challenge*, Population Reports, Series M, No. 15, Vol. XXVIII, No. 3, Baltimore, Johns Hopkins School of Public Health, Population Information Program.

http://www.nic.in

Registrar General and Census Commissioner of India, (2001). *Provisional Population Totals*, Census of India, Paper 1 of 2001, New Delhi.

Srivastava, H. C. and Anjula Saraff, (2001). *Population and Environmental Health in India*, Paper presented in National Seminar on "The Impact of Environmental Degradation on Human Population" in the New Millennium organised by R. S. P. (P. G.) College, Jharia, Dhanbad, January 20-21, 2001.

United Nations, (2001). *World Population 2000*, United Nations, Population Division, Department of Economic and Social Affairs, New York.

UNDP, (1998), *Unequal Impacts of Environment Damage*, Human Development Report 1998, Oxford University Press, New York.

UNDP, (2000). *Human Development Report*, United Nations Development Programme, Oxford University Press, New York.

WHO, (1992). *Our Planet, Our Earth*, Report of the WHO Commission on Health and Environment, World Health Organization, Geneva.

WHO, (1997). *Health and Environment in Sustainable Development: Five Years after Earth Summit*, Geneva.

7

Policy for Creating Science Literacy for Sustainable Development: A Case of Gujarat

NAROTTAM SAHOO and VINU T. J.

Introduction

The popularization of science and technology is intended to provide broad sectors of the population with the challenge and satisfaction of understanding the universe in which we live and, above all, being able to imagine and build possible new worlds.

Today, there are societies that produce knowledge, and those that are essentially consumers of it; there are societies that progress, build and create, and others that passively contemplate such progress, with little chance of understanding and adapting to the changes that progress implies.

Science and Technology have become crucial in determining the character of the society. We need a curriculum that is relevant, which is not detached from man's everyday ordinary experiences. Social issues that a student come across need an increasing degree of scientific and technological literacy for both the understanding and decision making to stimulate necessary action.

The areas to be reached by science and technology must be broadened to incorporate formal education and communication with non-formal ones, academic discourse with colloquial language, laboratory materials with domestic objects and ordinary daily achievements. It is also essential to make scientific and technological knowledge available to the ordinary citizen, which would allow for the making of informed opinions on these matters, so that they form part of daily conversations in the same way that politics and sport do.

Activities leading to the popularization of science and technology must, therefore, be based on interdisciplinary dialogue and work, integrating diverse fields of knowledge and different theoretical and methodological approaches. In the long run, the popularization of science and technology, as every other social and cultural activity, must necessarily have an impact on the economic and social development of nations, specifically on sustainable development and the quality of life; preservation of the environment; knowledge and strengthening of the national culture; the learning of ethical values; education that is objective, creative, participative, independent, impartial, and plural; and the conscious pursuit of excellence.

The Gujarat Science City

The Gujarat Science City is emerging as a large-scale science popularization platform for the enhancement of scientific temper and science literacy in the State. Based on the principle of 'edutainment'—education through entertainment, the Gujarat Science City is envisioned to be an impact making science appreciation and entertainment experience for the common citizen.

All the programmes of the Science City are intended to enliven the imagination, foster creativity and develop a spirit of inquiry, especially in young minds. It will be a platform where, visitors can discover the wonders of science and technology and have access to the most exciting and contemporary form of entertainment regardless of the social stratum, education or age group and creates a culture of learning.

The objectives of the Gujarat Science City are:

1. To develop a scientific temper,
2. Nurture and stimulate scientific faculties,
3. Promote innovative and experimental activities,
4. Showcase breakthroughs in Science & Technology, and
5. Establish a countrywide network.

The Gujarat Science City is unique in several aspects. It goes beyond creating a museum or a theme park and will have the right blend of fun and education to ensure long and repeated visits. Enjoyment and learning go hand in hand and the facilities are there to offer a range of experiences for families, children in school group, and other visitors. Although hands-on and minds-on exhibits have been the hallmark of the science city, increasingly they have been joined by whole-body and emersion experiences, such as simulators and large format films.

The Science City is being developed gradually with its physical facilities on a very beautiful stretch of land at the outskirt of Ahmedabad. Its location is in neighbourhood of lush green agricultural area with typical rural landscape. On one of its side, it has small village while on other side the suburban urban area reflects technological atmosphere. It has been conceptualized on hexagonal grid patterns that house various theme pavilions.

The Gujarat Science City is emerging with India's first IMAX 3D Theatre, Space and Communication Pavilion, Energy Education Park, Basic Science Gallery, Children Activity Centre, Musical Fountain, Amphitheatre, Planet Earth Pavilion, Life Science Park, Ride Simulator, Earth Station and basic amenities like parking and cafeteria amid the beautiful landscape of the campus.

The pavilions have indoor and outdoor exhibits and are linked through internal transportation network. Several exhibits have been fabricated at Kolkata, Delhi, Pilani, Baroda, Ahmedabad, Hyderabad and Bangalore with the direct involvement of various scientists and technologists of the country. Easy accessibility of the pavilions has been considered as the most important criteria as the site is large in size and weightage has been given to the minimum walking distance.

Science Literacy Concept at Gujarat Science City

Scientific literacy is essential for the intellectual, social and economical development of any society. In spite of India, being one of the largest producers and supplier of human resources in various branches of science, there is ample evidence of an all-pervading lack of scientific thinking, skills and temper all around us. Scientific literacy represents cognitive capabilities for utilizing science and technology information in human affairs for social and economic progress[1].

According to Eduardo Martinez, who expressed his concern about science literacy at the UNESCO World Conference on Science, Budapest, 1999: "the popularization of science and technology is intended to provide broad sectors of the population with the challenge and satisfaction of understanding the universe in which we live and, above all, being able to imagine and build possible new worlds. One of the major challenges facing developing countries is to make science and technology an essential part of the culture of the people."

Perwitt presents a more comprehensive conception of scientific literacy. He states that scientific literacy originates in the interactions of science and society and an individual—although not a scientist or technologist—who is willing and able to cope with science related, decisions both at the personal and societal level. His view embraces cognitive, affective and behavioural domains.

Hetherington and Hetherington [BSCS and Social Science Education Consortium SSEC] offered a literacy triad that included—

(i) *Practical literacy*—consisting of knowledge and skills needed to maintain our way of life.
(ii) *Civil literacy*—dealing with the meaningful participation in the democratic processes of our highly scientific society.
(iii) *Cultural literacy*—dealing with the understanding of science as a major human achievement and an integral part of our generic culture.

Miller contended, "in a democratic society, the level of scientific literacy in the population has important implications for science policy decisions". He defined scientific literacy in the contemporary situation as consisting of three dimensions:

(i) An understanding of the norms and methods of science (nature of science).
(ii) An understanding of the key scientific terms and concepts (i.e., science content knowledge).
(iii) An awareness and understanding of the impact of science and technology in society.

The above discussion indicates that social responsibility has an established life space in the scientific literacy literature.

Gujarat Science City is working towards the presenting the images of science and technology as a major component of the cultural landscape. It has a strong mission for public understanding of science and will be rich resources for learning and fun. It welcomes diversity as it has estimated annual visitors of 1.5 million. In addition to on site programmes and activities, it has expanding the reach though variety of out-reach programmes and strong ties to the community, membership programmes. Aiming to be an integral part of the educational infrastructure, the Gujarat Science City is catering to large numbers of school children and to nurture youth in and out the school.

Gujarat Science City celebrates scientific activities for students, teachers and the common visitors round the year. It is also able to create and sustain a culture of learning that fosters the process of personal inquiry, experimentation, communication, understanding, and the sharing of values about our world.

The popularization of science and technology must make such knowledge a central component of culture, social awareness and collective intelligence. It must also contribute to an effective integration of cultural, ethnic, linguistic, social and economic issues[2].

Science Popularization for Sustainable Development

Sustainable development seeks to reconcile environmental protection and development; it means nothing more than using resources no faster than they can regenerate themselves, and releasing pollutants to no greater extent than natural resources can assimilate them. For achieving the goal, science literacy plays a major role.

Science and technology popularization activities must contribute to the arduous and difficult task of separating transcendental knowledge and information from that which is banal, superficial, ephemeral and unnecessary. In the long run, the popularization of science and technology, as every social and cultural activity, must necessarily have an impact on the economic and social development of nations, more specifically:

- The sustainable development of a nation and the good quality of life of the population
- The preservation of environment
- The knowledge and strengthening of the national culture
- The learning of the highest ethical values
- An education that is objective, creative, participatory, independent, impartial and laic
- The consciousness and pursuit of excellence.

The achievement of the above goals require the implementation of strategies that would generate continuous and permanent spaces for education, information, debate, appropriation and construction of knowledge. Such strategies must be adapted to each segment of society. In order to develop and strengthen a dynamic scientific and technological culture, dissemination strategies must mobilize political, institutional, social and economic structure. This will enable the population to understand the complexity and "globalness" of the contemporary world, to develop competence that is transferable to the world of labour and production, to the field of study, artistic expression and daily life[3].

Gujarat Science City has envisioned the concern for creating a sustainable development through Science, Technology and Society (STS) approach and is designing its activities and programmes to attaining the goal in this direction. This will focus to the application of scientific knowledge and processes to social issues. STS is committed to the educability of all youth and enunciates its goals broadly as science education and citizenship education imperatives. That learning science in an STS context results in students with more sophisticate concept and ability to use process, skills, attitude towards science, use of science concepts and processes in their daily lives and in responsible personal decision making.

STS means determining and experiencing ways that these basic ideas and skills can be observed in society. Focusing on real-world problems through STS activities, which have science and technology components from student's perspectives, instead of starting with concepts and processes. This allows students to investigate, analyse, and apply concepts and processes to real situations.

Activities leading to the popularization of science and technology are being planned on the basis of interdisciplinary dialogue and work, integrating diverse fields of knowledge and different theoretical and methodological approaches. GCSC believes that the popularization of science and technology is a process of communicating and appropriating scientific and technological knowledge aimed at broad sectors of the population. It is similar to a pyramid, which rests on four components:

- Science and technology interactive activities;
- Multimedia programmes for science and technology popularization;

- Mass media and the dissemination of science and technology; and
- Formal education/science learning.

The popularization of science should be taken care of by scientists, communicators, educators, and artists who have a sound knowledge about the content of the scientific message, and those who know how to make its transmission and learning more efficient and entertainment way.

It seems to us that scientific societies are well positioned to help bring about a culture supportive of science and technology driven development through public initiatives and other measures. Representing a concentration of expertise and intellectual resources, they are in a position to take an equity stake in their countries' futures. Although a lack of funds is one reason why such societies have not maximized their potential, this can be more than made up for by partnerships with government organizations and private corporations, as well as by tapping the large work-forces of the country. This would have the effect of strengthening existing institutions and enhancing their stature[4].

More importantly, scientific societies can be established without the need for institutional infrastructure, official sanction or massive funding. It just requires the coming together of a core group of scientists committed to translating a vision into reality.

Sustainable Development: A Collective Concern

1. There is a close correlation between the degree of development attained by any country and the level of scientific and technological capability, which that country has built up. This correlation is becoming more and more evident. However, there is also an increased concern about the consequences of the applications of certain scientific and technological advances upon the natural environment and the cultural heritage of mankind.
2. Consequently one should not only aim at further development through science and its application, but more wisely at what one should describe and is universally conceived, especially after the 1992 Summit at Rio de Janeiro and 2002 Summit at Johannesburg—as sustainable development.
3. In order to participate in the sustainable development as thus

conceived, no small state can feel exempt from the need to build up its own indigenous scientific and technological base. One cannot rely on the importation of science and technology originating from foreign sources, for without such a base no real technology transfer can occur.

4. Delays in the growth of scientific and technological culture, can only mean retarded development, and this is to be seen against the background of an acceleration in such fields as genetic engineering, microelectronics, new materials and alternative energy. To secure our future prosperity we can not wait for a natural organic growth in our scientific awareness and proficiency; we have to spearhead a rapid and strategic science and technology penetration in all areas.
5. In June 1999, the World Conference on Science in Budapest identified the key role played by the use of scientific knowledge in alleviating poverty. The meeting produced a 'Declaration on Science' in which participants noted that greater use of scientific knowledge is essential in creating a better quality of life and a sustainable and healthy environment for present and future generations (UNESCO, 2000).

GCSC's Approach of Science Propagation

- *Conceptual, but numerate:* Scientific literacy calls for understanding, not for calculation ability. Even aspiring scientists should begin with non-techincal courses, and certainly non-scientists have little techniques, but an urgent need to integrate into their lives. Briefly, liberal-arts science should be conceptual and numerate but not algebraic.
- *Interactive:* At least since Socrates, good teachers have known that teaching by dialogue works better than teaching by telling. A convincing body of evidence shows that concept-based interactive engagement methods are far more effective than straight lecturing in teaching science. Interactive methods have certainly worked for stimulating active involvement by students even in classes of over 200. These techniques, by encouraging less motivated students to think for themselves and less confident students to express themselves, work especially well with non-scientists. There are many interactive techniques: brainstorming, provocative

discussion questions and verbal questions to the class, are planned for each exhibit and activities of the science city.

- *Trim the details and unify:* GCSC believes that scientific methodology should be a theme of every introductory course.
- *Modern:* Students and other interest persons deserve to know the best current description of the natural world. We are venturing into an era, where technology is no longer a limitation to bringing magic on the screen, only imagination. Older theories can be enlightening, but the focus in science literacy should be on current understanding. Pedagogical inertia is one reason we do not update in our formal courses.
- *Social:* Industrial democracy cannot survive unless citizens are literate about science related societal issues. If citizens cannot vote intelligently on these matters, then democracy will succumb either to gross misuse of technology or to a non-elected scientific elite that will make these decisions. The common people have little need for scientific techniques, but an urgent need to integrate science into their lives.

Conclusions

The areas to be reached by science and technology must be broadened to integrate formal education and communication with informal efforts in both fields, academic discourse with colloquial language, and laboratory materials with domestic objects and ordinary daily achievements. It is also essential to make scientific and technological knowledge available to the ordinary citizen, in order to allow him or her to form opinions on such matters that can form part of daily conversations in the same way as politics and sport.

Gujarat Science City focuses its activities into three directions: the general public, children and young adults, apart from educational institutions and organisations. Of course, efforts to make scientific and technological knowledge available and familiar to the general population should be supplemented by educational reforms that promote the role of science teaching in primary and secondary education, as well as a significant expansion in hands-on and minds-on experimental activities; science is, after all, an experimental endeavour.

Activities aimed at the popularization of science and technology point in various directions, from the 'distribution' of information in the mass media to formal education, and from the democratizing function of science and technology interactive centres to the creation of diverse spaces or processes for non-formal participation.

When nations are changing, or want to change, science can offer answers to problems of health, food, poverty, shelter and culture. Hence, scientific and technological literacy is mandatory for achieving sustainable development in the twenty-first century. To enable the common citizen to be scientifically savvy and to achieve the important contemporary goal of scientific literacy, the Gujarat Science City is all set to deliver a new recipe of education along with entertainment. The activities and programmes set the Science City will go on to fulfilling its numerous objectives and serve as a source of education through entertainment for years to come.

REFERENCES

1. Eduardo Martinez; *Boosting public understanding of science and technology in developing countries*; UNESCO Office, Montevideo, Uruguay.
2. *ASTC Sourcebook of Science*; Center Statistics, 2001.
3. Sudipta Ghose; *Hastening Scientific Literacy for All: Our survival depends on it*: Proceedings of the CASTME-UNESCO-HBCSE International Conference on Science, Technology & Mathematics Education for Human Development, Vol. II, pp. 49-53, 2001.
4. Leo Tan Wee Hin and R. Subramaniam; "How scientific societies can build better nations", *Nature*, 17 June 1999, Vol 399, p. 633.

II

Man and Environment: Sustainability Issues

8

Dynamics of Man-Environment Relationship in a Tribal Area: A Case Study of Maharashtra

NARINDER KAUR and SUDHA SRIVASTAVA

The present study attempts to examine one of the tribal regions, i.e. Talasari taluka of Thane district in Maharashtra. The general terrain of the taluka is hilly and dissected, and inaccessibility encouraged marginalisation of the area. The region is served by perennial rivers, and a number of non-perennial streams, with a large fluctuation in volume of water according to the season. The inhabitants live either in '*wadis*' consisting of groups of houses numbering 20 to 50, separated from each other by a couple of km. or in '*padas*' which are very small groups of houses, just 2 to 5 in number scattered in the hilly areas. Industries are, by and large, absent from the regional scene. The location (that includes non-proximity to metropolitan Mumbai), settlement pattern, level of socio-economic development, attitude of the inhabitants towards literacy and female education, and lack of transport facilities due to inaccessible terrain, have been some of the factors responsible for the underdevelopment of the region.

Through cartographic and statistical analysis, the physical and socio-economic characteristics of the region will be focused upon, the spatio-temporal changes that have occurred and the interrelationships between selected parameters of the environment.

Introduction

India is a land of diversity, not only as regards to the physical attributes but also with reference to its population characteristics. In a pluralistic Indian society, among the many communities that co-exist, some have been living in hills and forests in varying concentrations. These are known as "tribes". They are an indigenous group of people living close to nature in a symbiotic relationship with it. They can be differentiated from the non-tribes on the basis of certain characteristics of habitat and traits like language, beliefs, rituals, customs and institutions. Geographically their population is found scattered in many parts of the country, but they exhibit concentrations in three broad zones, viz., northeastern hills and forested areas, hills and forest stretches of Central India, and in the island chains especially of the Andaman and Nicobar group. They occupied relatively inaccessible regions, which resulted in their segregation from the mainstream. With greater interaction due to advances in technology in transport and communication sectors, there have been changes in certain aspects of their economy, while at the same time retaining their overall way of life.

In order to understand the changing land use/land cover and changing relationships with nature, one such area has been taken up for the study. The taluka of Talasari represents a typical tribal scenario in an otherwise urban district of Thane in Maharashtra. Talasari is situated in the northernmost part of Thane district, bounded by the Dangs of Gujarat in the north, Union territories of Dadra and Nagar Haveli to the east, Dahanu to the south and Arabian Sea to the west. It is 125 km. from the District Headquarters (Map 8.1). More than 90 per cent of the total population of the taluka is tribal. It is pertinent to note that there is no notified town area, and most of the working population is engaged in primary activities.

Methodology and Data Base

- Cartographic analysis is based on Survey of India topographical sheets on a scale of 1:50,000 and land use maps based on satellite imageries of 1990s.
- Statistical analysis is based on data from the District Census Handbook of Thane District, 1971 and 1991.
- Using the MapInfo GIS package maps have been prepared to show the pattern that emerges for land use/land cover and for the main occupations, i.e. cultivation and agricultural labour.

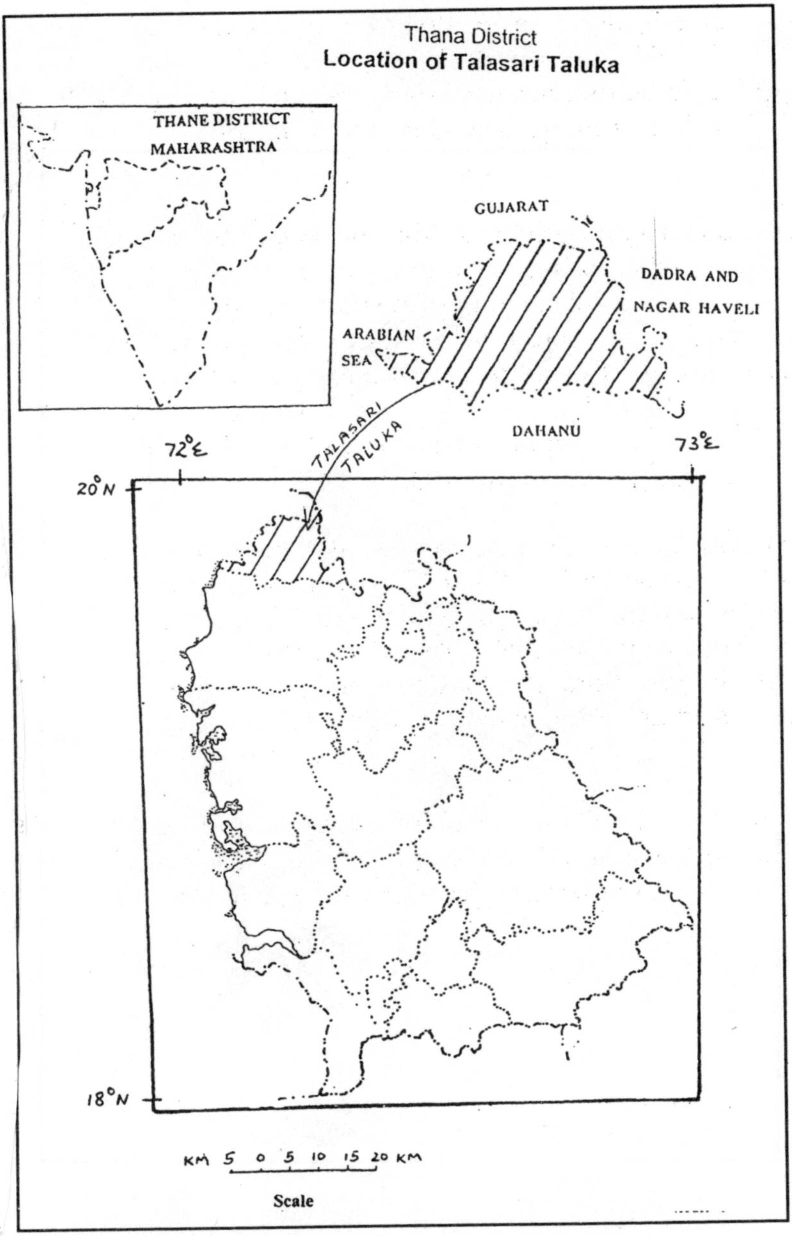

Figure. 8.1: Thana District: Location of Talasari Taluka

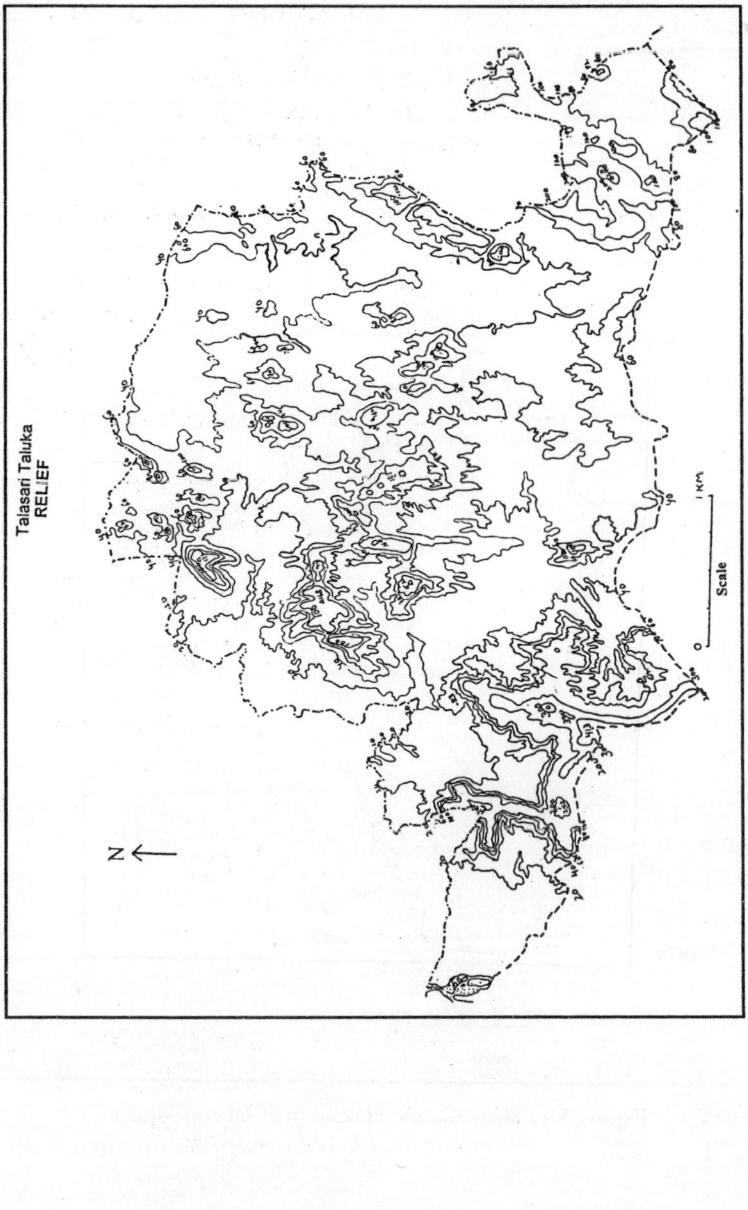

Figure 8.2: Talasari Taluka: Relief

Figure 8.3: Talasari Taluka: Drainage

The Study Area—Physical Setting

Talasari has a varied relief. The western part includes a limited section of the Maharashtra coast, the area being low-lying, gradually sloping towards the Arabian Sea. The district comprises of 41 villages covering an area of 266.83 sq. km (Map 8.1). The general appearance of the coast is submergent, and mudflats, rocks and sands are conspicuous. The rest of the region has a hilly terrain (Map 8.2), and as extreme erosion takes place, there is great variation in relief from the coast to the hills, the surface being undulating and irregular. The landscape changes from the coast to the interior as the surface rises rapidly. The maximum height of the taluka, i.e. 413 m is observed in the southwest. In the east, the maximum height is 286 metres. Geologically the region comprises the Deccan Trap with inter-trappean beds. The soil is skeletal, the depth is shallow and texture clayey or loamy.

The region is drained by three perennial rivers (Map 8.3), namely Juhi nadi in the extreme west bounding the taluka, Jogan nadi flowing from southeast to northwest and the Darota nadi in the northeast. There are a number of small streams, many of which are non-perennial, the water fluctuating substantially according to the season.

The climate is characterised by annual rainfall of 1759 mm during the monsoon months from June to October. The average maximum and minimum temperatures are 30.4°C and 22.3°C (Table 8.1).

Table 8.1: Talasari Taluka (Thane District): Temperature and Rainfall

Months	*Mean Daily Maximum Temperature (in deg. C.)*	*Mean Daily Minimum Temperature (in deg. C)*	*Rainfall (in mm)*
January	27.7	16.8	2.3
February	28.2	17.5	1
March	30.3	21	0.5
April	32	23.9	2.8
May	32.9	26.8	12.2
June	32.1	26.4	357.9
July	29.7	25.1	730.3
August	29.1	24.8	347
September	29.6	24.3	226.3
October	31.7	23	57.1
November	31.9	20	20.1
December	29.7	17.7	1.5
Annual Average	30.4	22.3	1759

Land Use/Land Cover

A considerable proportion of the land was originally under forest cover. The forests include:

(i) Reserved Forests that comprise dense mixed jungles and open jungles (which may contain teak),
(ii) Private forests which may be dense or open jungles.

Today less than 30 per cent of the area of the taluka is under forest cover (Map 8.4). This figure is even lesser than that for the entire district, i.e. 37 per cent. As per India's National Forest Policy, at least 20 per cent of lowland areas and 66 per cent of hilly terrain (with an average of 33 per cent for an entire region) should be forested in order to maintain environmental sustainability and stability both in uphill regions and the plains below. The recorded forest cover does not show any significant change since 1971 (Table 8.2). It is observed that the forest cover in Achchad, Kajail, Girgaon and Karajgaon is high, i.e. above 50 per cent of the geographical area in 1971 as well as in 1991. All these villages have moderate elevation and a moderate slope. Some of the villages like Upalat and Bormal towards the north have a forest cover between 33 per cent and 48 per cent. The land here has a low gradient. Most of the villages have less than 33 per cent of forest cover Maps 8.2, 8.3 and 8.4 here.

By overlaying maps showing physical attributes, analysis was done and the results have been tabulated in Table 8.3. Since it is an area of uneven terrain dominated by a tribal population one would expect a dominance of forests in the region. However, it is seen that cultivation, often accompanied by plantations/natural tree cover of different varieties, and forests are equal in dominance as far as area coverage is concerned. Scrubs are widely distributed in the area too. These are gradually taken over for cultivation or other development purposes. The staple crops are rice and jowar. While horticulture is found along coastal regions, the northeastern part has stretches of grassland (Tables 8.2 and 8.3).

It is clearly seen that forest is the main type of land use/land cover at elevations of 60-100 m, where the relative relief (i.e. difference between maximum and minimum heights which is indicative of steepness of slope) is between 40 to 60 m. Even within these Reserved Forests (under the Forest Department) are found at higher elevations and along steeper slopes, as compared to private forests. The latter

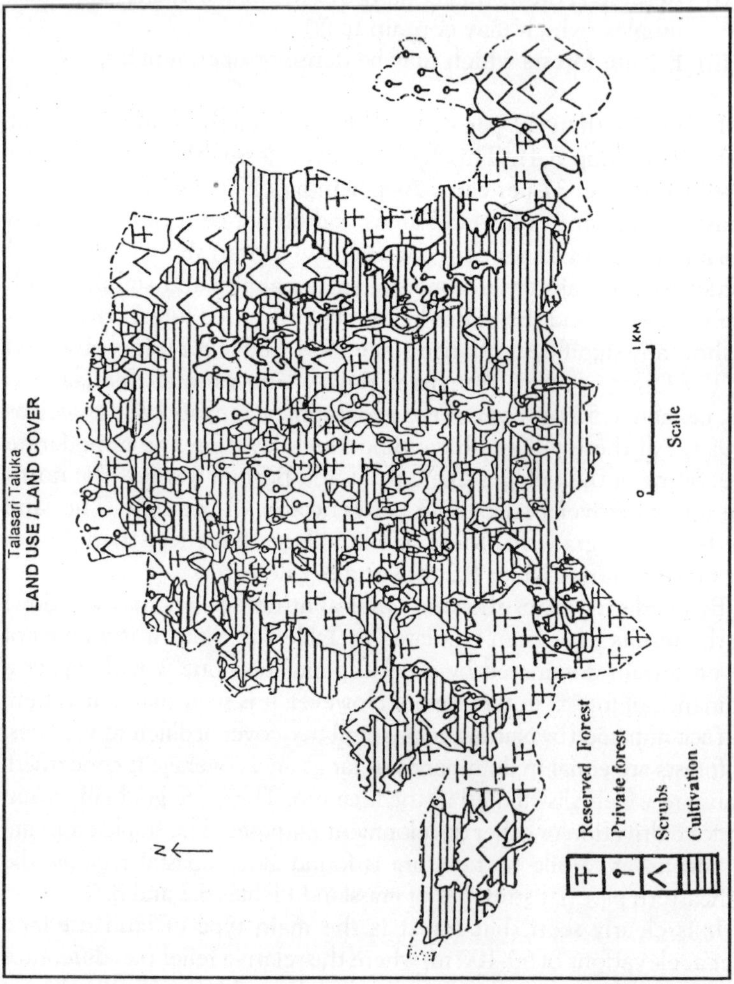

Figure 8.4: Talasari Taluka: Land Use/Land Cover

Table 8.2: Talasari Taluka (Thane District): Area under Forest Cover in Ha (1971, 1991)

Villages	*Forest (1971)*	*Forest (1971 %)*	*Forest (1991)*	*Forest (1991 %)*	*Difference %*
Sambha	261.22	43.4	261.06	43.54	0.14
Achchad	327.64	55.18	327.3	55.15	–0.03
Kajail	115.83	50.35	115.72	50.3	–0.05
Upalat	733.45	41.21	732.74	41.24	0.03
Bormal	132.43	35.22	129.5	34.45	–0.77
Kochai	2.43	0.26	2.41	0.26	0
Anwir	191.56	27.68	191.4	27.66	–0.02
Sawroli	112.99	16.65	113.05	16.75	0.1
Varwade	181.44	17.71	181.26	17.73	0.02
Amagaon	55.89	9.85	55.85	4.15	–5.7
Dongari	323.19	26.95	336.8	28.15	1.2
Ghimniye	166.86	34.97	166.69	35.28	0.31
Girgaon	745.6	51.68	790.51	55.82	4.14
Sutrakae	511.11	26.87	510.86	26.96	0.06
Zari	488.02	32.91	487.79	33	0.09
Vevaji	561.74	30.99	561.5	31.36	0.37
Borigaon	85.46	30.54	85.63	34.66	4.12
Zai	202.09	19.13	155.78	16.28	-2.85
Karajgaon	649.62	51.76	649.35	51.93	0.17
Vasa	328.46	28.75	328.04	29.04	0.29
Kawade	59.54	8.49	59.47	8.53	0.04
Sawane	62.78	12.22	63.11	12.29	0.07
Vadavali	163.22	21.28	161.32	22.01	0.73
Talasari	454.41	21.94	453.74	22.06	0.12
Kurze	456.84	33.66	456.56	33.87	0.21
Kodad	53.06	17.28	53.2	17.5	0.22
Udhawa	539.87	31	539.1	31.21	0.21
Talasari	7966.75	29.51	7969.74	29.87	0.36

are absent above 100 m and where the relative relief goes beyond 60 m. On very steep slopes (where relative relief is more than 80 m up to 280 m) and at heights above 140 m Reserved Forest is probably the only type of land cover found.

Cultivation predominates at elevation ranging between 20 and 40 m, but is constrained by steep gradients, hence is important only where relative relief is not greater than 10 m. It is not found above 100 m and where relative relief goes beyond 60 m.

The total agricultural area shows a slight decline over a twenty-year period. (Table 8.4). Within the taluka, the total agricultural area was 62.04 per cent in 1971, which has decreased to 59.63 per cent in 1991.

Table 8.3: Talsari Taluka (Thane District): Relation between Land Use/Land Cover, Elevation and Relief

Elevtion in (m)	*0-20*	*20-40*	*40-60*	*60-80*	*80-100*	*100-120*	*120-140*	*140-160*	*160-180*	*180-200*	*200-220*	*220-240*	*240-260*	*260-280*	*280-300*	*300-320*	*320-340*			
Reserved Forrest		2	1	2	13	2	3	5	2	3	2	1			4		2	42		
Private Forest	1	3	7	4														15		
Cultivation	5	16	20	10	4	1		1					1					58		
Scrubs	3	15	9	7	9	1												44		

Relative Relif in (m)	*0*	*10*	*20*	*30*	*40*	*50*	*60*	*70*	*80*	*90*	*100*	*120*	*140*	*160*	*>200*					
Reserved Forest		5	3		7		12		1		5	2	1	2	4			42		
Private Forest		4	4	0	6	0	1											15		
Cultivation		33	12		6		6							1				58		
Scrubs		21	16		4		2		1									44		

Table 8.4: Talsari Taluka (Thane District): Area under Agriculture in Ha (1971, 1991)

Villages	*Agricultural Area (1971)*	*Agricultural Area (1971 %)*	*Agricultural Area (1991)*	*Agricultural Area (1991) %*	*Difference (%)*
Sambha	304.16	50.54	275.59	45.9	–4.64
Achchad	233.28	29.29	233.09	39.2	–0.09
Kajail	95.18	41.38	95.1	41.3	–0.08
Upalat	982.53	55.2	981.66	55.2	0
Bormal	218.29	58.08	218.12	58	–0.08
Kochai	782.87	83.18	782.25	83.27	0.09
Anwir	403.38	58.28	403.06	58.26	–0.02
Sawroli	471.82	69.51	471.46	69.84	0.33
Varwade	743.99	72.64	744.61	72.83	0.19
Amagaon	460.08	81.09	459.54	81.16	0.07
Dongari	746.82	62.28	263.04	21.99	–40.29
Ghimniye	266.49	55.86	183.96	38.94	–16.29
Girgaon	575.1	39.86	548.18	38.71	–1.15
Sutrakae	1239.71	65.17	1238.74	65.3	0.13
Zari	869.54	58.63	962.41	65.12	6.49
Vevaji	1119.02	61.74	1097.13	61.28	–0.46
Borigaon	180.63	64.55	147.67	58.77	–4.78
Zai	758.16	71.48	643.23	67.26	–4.52
Karajgaon	491.27	39.14	531.04	42.47	3.33
Vasa	713.61	62.46	713.05	63.1	0.64
Kawade	574.69	81.93	568.25	81.55	–0.38
Sawane	409.05	79.65	408.73	79.6	–0.05
Vadavali	554.85	72.33	554.09	75.6	3.27
Talasari	1429.25	69.02	1265.21	61.66	–7.36
Kurze	830.25	61.18	829.57	61.5	0.32
Kodad	223.16	72.69	229.99	73.35	0.66
Udhawa	1070.42	61.47	1059.75	61.35	–0.12
Talasari	16747.6	62.04	15911.52	59.63	–2.41

Out of 27 villages, the agricultural area of 14 villages exceeds the average for the taluka in 1991 and in the remaining 13, it is below the taluka average as shown in Table 8.4. In almost all the villages, irrigation is done by wells. There is hardly any change in the agricultural area of most villages between 1971 and 1991. However, Dongri shows a substantial decline in agricultural area from 1971 to 1991 (40.29%). Ghimniye too shows a decline in agricultural area from 55.86 per cent in 1971 to 38.94 per cent in 1991 amounting to 16.92 per cent.

Scrubs are found at heights similar to those where cultivation dominates, i.e. 20 m to 40 m, and where slopes are not steep. They are excellent indicators of deforestation in areas where environmental conditions are conducive to good forest cover. The fact that they abound in areas that are transitional between cultivation and forests suggests that they are in a stage between the two. They occupy areas that are accessible but not easily cultivable. The constraints are elevation and slope. Nevertheless, with increasing pressure for cultivable land, they become the most vulnerable. Thus they represent a stage in retrogression from a forest cover. This is amply demonstrated by the fact that they are absent above 120 m and where relative relief is above 80 m. Thus increasing inaccessibility has acted as deterrent, and helped in maintaining a much needed forest cover at higher elevations.

Demographic Characteristics

The total population of the taluka is 92955, showing a growth rate of 74.12 per cent since 1971 (Table 8.5). Achchad, Vasa, Sawane and Vadavali recorded a growth rate of more than 100 per cent, while Borigaon and to some extent Amagaon recorded a growth rate much lower than the average for the taluka. Karajgaon is the only village to record a negative growth.

It is clear from Table 8.6 that Talasari is a tribal region. It shows the village-wise distribution of scheduled tribes and percentage of scheduled tribes to the total population for the years 1971 and 1991. From 88.31 per cent in 1971 the proportion of scheduled tribes has increased to 91.53 per cent along with the increase in total population. The growth rate of the tribal population is greater than that for the total population, i.e. 80.48 per cent. In most cases, the growth rate of total population and of tribal population is comparable, but a few exceptions stand out. Borigaon has a total growth rate of merely 9.6 per cent, while the growth rate of tribal population is 232.21 per cent! In Karajgaon, while there is a decline in total and tribal population, the proportion of tribal population has shown an increase to 100 per cent. Sawroli, Vasa, Sawane, and Vevaji show a high growth rate for tribal population but a much lower one for the total population.

The density of population has correspondingly shown a remarkable increase (Table 8.6). The change in the pattern is clearly seen in the maps (Map 8.5). In the year 1971 the overall density of population in the taluka was 208 persons per sq. km which has increased to 348 persons per sq. km in 1991.Among the 27 villages Sambha shows the

Table 8.5: Talasari Taluka (Thame District): Demographic Charactreistics (A)

S.No	Villages	Population 1971	Population 1971	Growth Rate (%)
1.	Sambha	1877	3291	75.33
2.	Achchad	846	1888	123.17
3.	Kajail	441	727	64.85
4.	Upalat	3564	6747	89.31
5.	Bormal	660	1016	53.94
6.	Kochai	1977	3239	63.83
7.	Anwir	1270	1965	54.72
8.	Sawroli	1117	1947	74.31
9.	Varwade	1749	3307	89.08
10.	Amagaon	1092	1333	22.07
11.	Dongari	2677	4194	76.44
12.	Ghimniye	1226	2323	89.48
13.	Girgaon	2568	3357	59.58
14.	Sutrakae	2983	5087	70.53
15.	Zari	2734	3718	63.02
16.	Vevaji	3922	1882	72.57
17.	Borigaon	479	525	9.6
18.	Zai	3300	3161	41
19.	Karajgaon	1320	939	28.86
20.	Vasa	1635	3667	124.28
21.	Kawade	1682	1892	76.75
22.	Sawane	732	1745	137.74
23.	Vadavali	1685	3759	123.09
24.	Talasari	5468	6195	81.71
25.	Kurze	2660	3565	34.02
26.	Kodad	815	1260	54.6
27.	Udhawa	3204	4572	69.35
	Talasari	53385	92955	74.12

highest density of population in 1991 followed by Vadavali, Ghimniye, Zai, Talasari and Kawade all have more than 500 persons per sq. km. In all these, villages density has increased along with the total population. The lowest density of population is found in Karajgaon, i.e. 189 persons per sq. km. The rest of the villages have shown increase in population but show moderate densities.

By comparing the data of 1971 and 1991, it has been found that the villages, which were having density of population less than 200 persons per sq. km, now have moderate density ranging between 200 to 400 persons per sq. km except Karajgaon, where density is still low probably due to the high altitude of the region. Zai which is a coastal area shows an increase in the density of populaton from 312 persons per sq. km to 486 persons per sq. km.

Table 8.6: Talsari Taluka (Thane District): Demographic Characteristics (B)

Villages	*Scheduled Tribes 1971*	*Total Population 1971*	*Scheduled Tribes 1971%*	*Scheduled Tribes 1991*	*Total Population 1991*	*Scheduled Tribes 1991%*	*Growth Rate ST*
Sambha	1534	1877	81.73	2717	3291	82.56	77.12
Achchad	287	846	97.75	1427	1888	75.58	72.55
Kajail	425	441	96.37	677	727	93.12	59.29
Upalat	3405	3564	95.54	6599	6747	97.81	93.8
Bormal	594	660	90.00	955	1016	94.00	60.77
Kochai	1961	1977	99.19	3213	3239	99.20	63.84
Anwir	1270	1270	100.00	1965	1965	100.00	54.72
Sawroli	925	117	82.81	1883	1947	96.71	103.56
Varwade	1735	1749	99.20	3244	3307	98.09	86.97
Amagaon	1081	1092	98.99	1311	1333	98.35	84.92
Dongari	2367	2377	99.58	4173	4194	99.50	76.29
Ghimniye	1219	1226	99.43	1970	2323	84.80	61.61
Girgaon	2474	2568	96.34	3231	3357	96.25	59.9
Sutrakae	2903	2983	97.32	4990	5087	98.09	71.89
Zari	2680	2734	98.02	3412	3718	99.84	65.52
Vevaji	2811	3922	71.67	1623	1882	86.21	65.52
Borigaon	149	479	31.11	495	525	94.29	232.21
Zai	2171	3300	65.79	1397	3161	44.19	32.19
Karajgaon	1295	1320	98.11	939	939	100.00	82.24
Vasa	1602	1635	97.98	3633	3667	99.07	126.78
Kawade	1650	1682	98.10	1793	1892	94.77	74.18
Sawane	734	734	100.00	1731	1745	99.20	135.83
Vadavali	1631	1685	96.80	3716	3759	98.86	127.84
Talasari	4265	5468	78.00	5617	6195	90.67	81.92
Kurze	2174	2660	81.73	3180	3565	89.20	46.27
Kodad	807	815	99.02	1255	1260	99.60	55.5
Udhawa	2453	3204	76.56	4324	4572	94.58	83.57
Talasari	47142	53385	88.31	85081	92955	91.53	80.48

For the entire taluka data reveals that 45 per cent of the population comprises the main workers (Table 8.7). Of these almost 68 per cent are cultivators and 13.7 per cent are agricultural labourers. Rest of the activities like livestock, mining, construction, trade and commerce does not seem to be an important activity in the area. These activities show a low percentage for the entire taluka. In some villages like Achchad, Sambha, Sawroli and Vevaji manufacturing seems to be an important activity. Vevaji shows high percentage, i.e. 89 per cent of its main workers engaged in manufacturing other than household while only 33 per cent and 6 per cent of the workers are cultivators and agricultural labourers respectively. In spite of a hilly

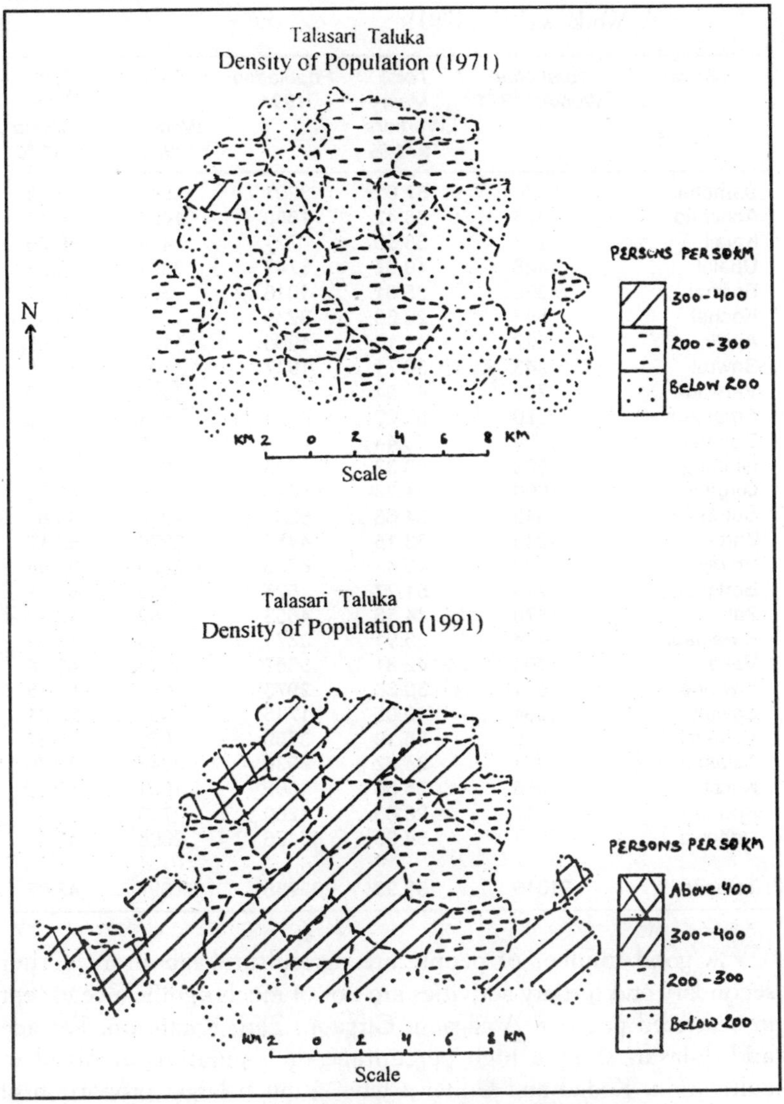

Figure 8.5: Talasari Taluka: Density of Population (1971)

terrain, the tribal population here is mainly engaged in agriculture, probably of the subsistence type. Forestry does not seem to be as important an activity as cultivation. Only in Vevaji, Borigaon, Zai and Gandhinagar forestry and related activities appear to be important occupations.

Table 8.7(a):Talasari Taluka (Thane District): Classification of Main Workers (1971, 1991)

Villages	*Total Main Workers1971*	*Total Main Workers 1971 %*	*Population 1991*	*Total Main Workers 1991*	*Total Man Workers 1991 %*
Sambha	1070	57.01	3291	1543	46.89
Achchad	325	38.42	1888	1027	54.40
Kajail	258	58.50	727	434	59.70
Upalat	1428	40.07	6747	3391	50.26
Bormal	235	35.61	1016	396	38.98
Kochai	587	29.69	3239	1283	39.61
Anwir	522	41.10	1965	1137	57.86
Sawroli	343	30.71	1947	735	37.75
Varwade	709	40.54	3307	1897	57.36
Amagaon	570	52.20	2021	821	.40.62
Dongari	1199	50.44	4194	1976	47.11
Ghimniye	568	46.33	2323	837	36.03
Girgaon	1060	41.28	4098	1709	41.70
Sutrakae	945	34.68	5087	2076	40.81
Zari	1043	38.15	4457	2370	53.17
Vevaji	1782	45.44	6768	2739	36.04
Borigaon	248	51.77	525	240	45.71
Zai	1470	44.55	4653	2182	46.89
Karajgaon	474	35.91	2371	1359	57.32
Vasa	700	42.81	3667	1703	46.44
Kawade	511	30.38	2973	1521	51.16
Sawane	294	40.05	1745	494	28.31
Vadavali	930	55.19	3759	1679	44.67
Talasari	1871	34.22	9936	3348	33.70
Kurze	866	35.56	3565	1901	53.:32
Kodad	217	26.63	1260	733	58.17
Udhawa	1091	34.05	5426	2666	49.13
Talasari	21316	39.93	92955	41897	45.07

A good number of people are agricultural labourers. Other secondary and tertiary activities are not of much significance except in scattered pockets. Amagaon, Girgaon, Zari, Karajgaon, Kawade and Talasari show a high percentage of population involved in cultivation. Kodad and Udhawa have a much larger proportion of population engaged in agriculture. Vevaji has a high percentage of population engaged in manufacturing, processing and services, in addition to agriculture. Zai has a high proportion of population engaged in fishing activities, as it is a coastal location. Many among them are also working as agricultural labourers.

Table 8.7 (b): Talsari Taluka Distirct: Classication of Main Workers (1971, 1991)

Villages	*A*	*B*	*C*	*D*	*E*	*F*	*G*	*H*	*I*	*J*
Sambha	504	47.1	386	25.02	–22.8	546	51.03	645	41.8	–9.23
Achchad	240	73.852	98	9.54	– 64.31	70	21.54	541	52.68	31.14
Kajail	105	40.7	8	1.84	–38.86	139	53.88	69	82.49	28.61
Upalat	244	17.09	54	1.59	–15.5	1166	81.65	2934	86.52	4.87
Bormal	46	19.57	2	0.51	–19.06	187	79.57	382	96.46	16.89
Kochai	78	13.29	78	6.08	–7.21	501	85.35	1079	84.1	–1.25
Anwir	256	49.04	64	5.63	–43.41	265	85.77	1001	88.04	37.25
Sawroli	143	41.69	78	10.61	–31.08	194	56.56	496	67..48	10.92
Varwade	131	18.48	192	10.12	–8.36	567	79.97	1523	80.28	0.31
Amagaon	205	35.96	110	13.4	–22.56	362	63.51	591	71.99	8.48
Dongari	242	20.18	305	15.44	–4.74	933	77.81	1203	60.88	–16.93
Ghimniye	344	60.58	305	36.44	–24.12	177	31.16	353	42.17	11.01
Girgaon	393	37.08	463	27.09	–9.99	556	52.45	757	44.29	–8.16
Sutrakae	115	12.17	346	16.67	4.5	799	84.55	1597	46.93	50.14
Zari	185	17.74	338	14.26	–3.48	832	79.77	1738	73.33	–6.44
Vevaji	470	26.37	157	6.44	–19.93	1251	70.2	1210	49.61	–20.59
Borigaon	174	70.16	123	51.25	–18.91	69	27.82	44	18.33	–9.49
Zai	906	61.63	492	22.55	–39.08	323	21.97	940	43.08	21.11
Karajgaon	22	4.64	601	44.22	39.58	448	94.51	706	51.95	–42.56
Vasa	152	21.71	219	12.86	–8.85	537	76.71	1262	74.1	–2.61
Kawade	34	6.65	367	24.13	17.48	455	89.04	1090	71.66	–17.38
Sawane	12	4.08	7	1.42	–2.66	281	95.58	452	91.5	–4.08
Vadavali	65	6.99	278	16.56	9.57	825	88.71	1174	69.92	–18.79
Talasari	262	14	237	7.08	–6.92	1038	55.78	2103	62.81	7.33
Kurze	44	5.08	215	11.31	6.23	598	69.05	1596	83.96	14.91
Kodad	3	1.38	30	4.09	2.71	211	97.24	693	94.54	–27
Udhawa	138	12.654	181	6.79	–5.86	735	67.37	1924	72.16	4.79
Talasari	5513	25.86	5734	13.69	–12.17	14065	65.98	28392	67.77	1.79

Conclusions and Suggestions

Talasari was once a part of Umbergaon (now in Gujarat) of the erstwhile Bombay Province. The National Highway and the Western Railway pass through the taluka in close proximity and this has changed the characteristics that are normally seen in the tribal talukas located in the interior. It is better developed and well connected, and this accounts for the dominance of agriculture and not forestry as the major activity. People here are definitely better off in terms of living conditions and quality of life. Moreover, social workers have contributed greatly to the betterment of the population. Industrial centres like Vapi and now Umbergaon have also had an impact. It is now being identified as a tourist place too. Its history coupled with its geography makes it very different from the neighbouring areas.

The total area under cultivation has shown a slight increase, though a clear decrease is recorded in Girgaon, Vevaji, Zai, Kawade, Talasari and Udhawa along with the new villages created from them. Substantial increase is noted in Amagaon, to a lesser extent in Zari, and Karajgaon. But forest cover in these regions has remained the same. It probably indicates that scrubs have been converted to cultivated lands.

Over a decade, agricultural area in these villages has increased by merely 2.15 per cent, while forested area has shown a marginal decrease of –0.04 per cent. 27 villages of the earlier census report have been re-aligned to constitute 41 villages. The new villages appear to have been formed on the basis of "area under irrigation". This is inferred from the fact that new villages have no irrigated area while the core village shows no change in irrigated area as seen in Talasari, which has been split into four new villages.

The presence of forests is essential for maintaining the environmental balance. Large areas have been taken over for cultivation. In the light of the present situation, it is suggested that

- Scrublands should be protected from further degradation,
- Scrublands should be gradually reverted to forestland,
- Core forest areas should be maintained as dense covers,
- Open forest areas should be improved in both quality and density by planting indigenous trees,
- Conversion to non-forest activities should be discouraged in hilly terrain,
- Felling of timber yielding trees should be controlled,
- Identification of areas for reforestation/afforestation programmes needs to be done on a priority basis,
- People's participation must be encouraged as they alone know the nuances of systems that they have been associated with for generations,
- Their traditional knowledge must be tapped for conservation of biotic resources, and
- Whatever development takes place must take care of the requirements of the local population and at the same time be within the framework of the environmental equilibrium.

Ecologically friendly man-environment relationships of the past may not fit into the scheme of things of the present world situation. Change is intrinsic to all dynamic systems and should not be seen as a 'negative' development. However, it is equally essential to conserve the stability of such interactive systems so as to ensure sustainability.

REFERENCES

Amrite, V. G. (1965). "Land Utilisation in Coastal Thana District (North Konkan)" in *Bombay Geographical Magazine*, Vol 13(1), pp. 149-160.

Bose, N. K. (1971). *Tribal life in India*, National Book Trust, New Delhi.

Chaudhary, B. (1982). *Tribal Development in India: Problems and Prospects*, Chugh Publications, Allahabad.

Dikshit, K. R. (1991). *Environment, Forest Ecology and Man in the Western Ghats—The Case Study of Mahabaleshwar Plateau*, Rawat Publication, New Delhi.

Government of Maharashtra (1981). Census of India, District Census Handbook of Thane.

Government of Maharashtra (1964). *Gazetteer of India*, Maharashtra State, Thane District.

Meher-Homji, V. M. (1987). "Vegetation Types of Peninsular India in Relation to Environmental Conditions", in *Explorations in the Tropics*, Prof. K. R. Dikshit Felicitation Volume, Pune, pp. 3-19.

Spate, O. H. K. and Learmonth, A. T. A. (1967). *India and Pakistan: Land, People and Economy*, Methuen & Co., London.

Srivastava, S. (1995). "The Phytogeography of North Konkan, with special reference to forest cover", University of Mumbai, unpublished thesis.

9

Ecological Resources and its Impact on Urban Areas

SURINDER SUNEJA

The Indian forests are rich in flora and fauna. The plant wealth of India is about 45,000 species constituting about 12 per cent of all global plants. There are 81,000 species of animal found in India. At present in India nearly 4.8 per cent of geographic area (1.5 million ha) and 18 per cent of good density of forest are being managed under the protective area network through 447 Sanctuaries and National Parks. All kinds of rights and privileges have been suspended in this area and by and large population support the endeavour of government though of late, lot of conflicts have emerged.

In India, out of the densely covered forest area of 38 million ha are under category dense forest (crown density 40% or above).

Uncontrolled and unsustainable removal (due to increased pressure) without adequate and commensurate investment leading to erosion in growing stock or qualitative decline in the productivity, change in land use on account of diversion of forest land to non-forestry uses resulting in disappearance of forest. In 1996, the wood budget of India was as stated below:

	Demand	*Supply*		*Gap*
		Forest	*Non-forest*	
Fuel and Wood (million tonnes)	201	17	98	86
Timber (million cubic metre)	64	12	31	21

Forest contribute to 1.7 per cent of GDP of the country on a moderate scale, the annual value of this is around Rs. 4,00,000 million (US $ 10,000 million) [Unpublished Research Report by Jigyasa, 2002].

To check the felling rate, the following procedures/techniques need to be incorporated:

Strategic Assessment for Conservation (SAC)

It is a comprehensive regional land use plan with emphasis on planning a system of conservation area in relation to human activities and constraints.

Methodology of SAC

(i) Establish a database using a series of "Compendia" such a species, the physical environment, the biotic environment, human activities and institutions.

(ii) Maps should be analysed for correlation or interaction between species ranges, biotic environment, physical and human feature, overlay techniques may be used in identifying area of potential conservation interest.

(iii) Areas of conservation interest may be ranked using criteria such as size, species of diversity, uniqueness, representativeness and other human criteria. These may be assigned to appropriate use categories such as park, nature reserve, game reserve, buffer zone and recreation areas etc.

(iv) Existing and potential human resource area interaction should be identified and mapped.

(v) General management and development options should be evaluated to minimize conflicts.

To arrive at logical and rational decision, conservationist must have good economic argument to save forest and wild life. This may also include population growth, recreation needs, tourism, pollution, forest cover, watershed benefits, etc.

Regional Environmental Management Plan

It has many objectives in common with strategic assessment above but an extended period of data collection and analysis is normally not feasible as the time deadline is critical. The experts must quickly map forest coverage, watershed management, etc. Necessary monitoring and surveys were written up as new projects including preparation of terms of reference (TOR). The TOR must be realistic and feasible.

Environmental Impact Assessment (EIA)

An environmental impact assessment (EIA) deals only with impact caused by a particular project, ideally it should involve the advance planning of forest and wildlife enhancement or mitigation measures which should be incorporated into the design of the project.

Survey of wildlife for EIA should consist of more than just listing of species that occur in the area. It should include mapping of distribution of habitat, estimates of relative abundance and samples of species diversity.

The lack of incorporation of forest and wildlife values in EIA is due to lack of methodology for evaluating of the benefits.

Conservation Management Plan

A management plan is prepared in order to aid the "orderly" development and effective management of conservation area such as wildlife reserve. Major benefits which accrue are:

(i) Establish the objectives and values of the protective area. This will foster beneficial development and prevent appropriate development in the area.
(ii) Establish manpower needs for protection and management.
(iii) Identify important research and monitoring needs.

Habitat and Resource Inventories

The procedure followed for the habitat and resource inventories are:

(i) Assemble all available references and maps, recent satellite images, vegetation, land use maps and topography maps will be very useful.
(ii) Procurement of 'Red Data Sheets', which have information on all endangered or vulnerable species that might occur in the area.
(iii) Plan the methods of inventory. Select a group of important

species for distribution and relative abundance survey. These species to be selected on the basis of:

(a) Degree of endangerness.
(b) Potential economic or social value including the interest to tourists.
(c) Ecological value to the community such as predator, key food species, etc.
(d) Indicator species, which reflects the "health" of a given habitat.
(e) Rapid methods of vegetation classification and assessment to be selected such as canopy closure and height, indicator trees species, cover of important tree species.
(f) Ground survey: It should be taken in the form of transect walk or establishment of sampling stations spaced regularly through the area.
(g) Sampling of diversity: Comparative estimates of the diversity of the selected tax as should be made from replicated samples of observations per unit area, transit length, time or other measures of sampling efforts.
(h) Comparison with survey results from other areas should be made to attempt to answer the questions, what is unique?
(i) The results should be summarized and evaluated in terms of monitory or non-monitory benefits, what is the value of the area in comparison with other areas in terms of uniqueness of fauna, flora and habitat, rarity of species present and the importance to the survival of endangered species?
(j) Population studies and census—it involves two phases:

   - Intensive study of behaviour and ecology at one particular site; and
   - Extensive survey to map the present distribution of the species and estimated population size.

(k) Base line studies and monitoring includes:

- Measurement of long term or gradual environmental impact of a development project.
- Measurement of a long term changes either beneficial or detrimental following implementation of regional environmental plan.
- Evaluation of the effectiveness of management action in a conservation area.
- Changes in the population size, age structure etc. of an endangered species.

A base lime study is necessary at the initiation of a monitoring programme to establish initial values for all variables being monitored. The base line measurement should also be the most intensive in order to be able to calibrate test and interpret measurement to be made by subsequent monitoring.

Soil Fertility

One of the environmental concerns about the great increase in use of the Commercial fertilizers (Cfs) since the early 1960s has been trend towards decreasing use of animal manures and other agricultural wastes. Over time this trend to deplete soil fertility. Review of literature including the reports over a period up to 50 years indicates that:

- Recycling of Soil Organic Matter (OM) through plugging under of rice, other crop residuals or by adding manure or compost to soil is essential for maintaining the physical/ chemical/micro-biological properties of soil on which soil fertility depends, and if there is no recycling of OM & Cfs, there will be gradual loss in fertility.
- However, because of the phenomena involved including wide diversity in soil types and properties, it is not yet possible to make definitive estimates.
- The optimal practice seems to be a combination of use of OM & Cfs but it is not possible to quantify the desirable property for a given situation.

Land Clearing

Land clearing is defined as the removal of trees/vegetation from the forest area (ecological resource), this also includes primary and secondary forested areas. For the purpose of creating sufficient land to permit establishment of economical farming operation along with activities, i.e. grazing and aquaculture.

Elements of Successful Land Clearing Project

Agricultural productivity of forest area—the initial requirements to designate an area as a forest require base line survey for determining the agricultural values of the cleared land for cropping and related agricultural activities, based on the following considerations:

- Soil fertility and other soil characteristics, compared to alternative crops or uses;
- Water availability (Rainfall or Irrigated);
- Erosion control problems (depending upon slope and soil characteristics);
- Economic and socio-economic analysis;
- The socio-economic analyses to show that the overall operation will be sufficiently cater to the needs of the farm-economic families;
- Optimize harvesting and marketing of the cut material; and
- Manual clearing methods to be encouraged to create local jobs.

Selection Parameters for Protected Area within Ecological Resources

The four protected qualities for protected area designation are:

- Nature conservation values, arising from ecological process and those induce by human activities;
- Landscape values including farming and settlement patterns;
- Cultural value such as life style and social traditions; and
- People-Nature relationship in harmonious manner.

The Value of Ecological Resources

Rapid urbanization has been the dominant feature in the growth of developing countries, simply because the economic opportunities are better in the urban as compared to rural sector. This trend can be expected to continue over the foreseeable feature.

The urbanization "crunch" has taken place despite the lack of infrastructure in the most urban centres; that is, lack of facilities to accommodate the in-migration. This phenomenon led to the development of "poor people" or slum areas within the urban complexes, causing strain on all types of urban services, including management of wastes from homes and industries including sanitary sewage and excreta.

The values of ecological resources are primarily concern with long-term benefits, which they offer to socio-economic development. Unfortunately such resources have little economic value.

Forest as an Ecological Resource in Urban Context

A forest in urban context exposed to following factors, i.e. *sunlight, heat balance, water balance, air quality, air and root space, stress and disturbances, soil quality*, etc.

Sunlight

Great variation in length of exposure to sun, some areas are in permanent shade. Trees often forced to adapt to location mis-matched to shade tolerance.

- Night lighting common, it is determinate to tree health.
- Smog/Haze reduces effectiveness of sunlight absorb blue-violet end of the spectrum.
- Lower leaves often receive substantial reflected light of paved areas and building.

Since light condition in urban environment vary widely, the light tolerance of tree is critical. For example, forests under story tree are better suited for shaded area. Trees that inhabit cool north facing slope may grow well in shade on buildings, while those preferring south facing sunny slopes may be appropriate for urban area.

Heat Balance

Relatively higher temperature increased in water stress. Trees are susceptible to over heating and bark burn. Temperature gradient much more variable, lower layers often hot due to reflection of paved areas complex vertical temperature gradients.

Trees with thicker bark and leaf cuticle are more resistance and heat damage and can grow in urban areas. The delicate species that inhabits cooler area including north facing slopes can grow in cool and protected environment.

Water Balance

- Paved surfaces drain away water resulting in overall water deficits. Reduce infiltration and percolation rates, local variations in non-homogeneous soil profile.
- Soil water nearly always contaminated with heavy metals, petroleum product, herbicides etc.

The water content of urban soil vary widely requiring soil moisture condition to be studied prior to tree selection. The trees, with their adapting root systems, can withstand water logging under some paved surface. Where as trees adapted to drier upland areas can be grown in well-drained lands and parks. Drought adapted trees and those native to a region can be grown successfully areas of permanent water deficits.

Air Quality

Air humidity considerably generally lower, dry period results in extensive leaf drainage.

- Dust and particulate laden reduce the ability to leaves to transpire effectively.
- Ever present smog and chemical pollution are deleterious to tree health.

The solution to this problem is, of course, to reduce the source of urban air pollution in the forest phase. Trees with smooth, shining leave trap less particulate matter than those with hairy leaves.

Air and Root Space

Horizontal space varies from unlimited (plazas and parks) to severely constrain by vertical surfaces.

- Clearance for overhead wires severely limit vertical growth, provision for movement corridors and sightlines limits growth of lower branches.
- Under ground structure and utilities constricts overall rooting space. Hardy trees that survive in rocky, low soil nutrient conditions in the world are well suited for urban conditions. Flood plain trees that have adapted to alkaline soil may be suited to alkaline urban soil.

Stress and Disturbance

Absence of litter/predominance of grass in park lessen chances of potential fire sources leading to conflagration.

- Predominance of tall buildings reduce chances of trees being struck by lightning.
- Building configuration often shelter trees from wind conversely, they channel winds to higher velocities, single tree less subject attrition.
- Paved areas above root zone provide structural support. Lessen chances of wind throw. Poor soil conditions can result in under developed root systems increased chances of wind throne.
- Single trees face less or no competition.
- Lower number of wild animals less damage from herbivores. Dog urine in tree pit source of soil pollution adversely affecting soil chemistry.
- Frequent physical damage due to construction, grading pedestrian and vehicular circulation, recreation. Further damage due to careless grading and pruning.

Soil Quality

- Organic material removed, short circulating the nutrient recycling processes, soil is deficient in nitrogen.
- Difficult to work.
- Poorly structured.
- Varied colour including gray (a clay like soil that indicate poor aeration within the soil layers).
- Reduced micro organism activity.
- Results in limited break down existing organic matter with subsequent reduced availability of nutrient to plants.
- Minimal water movement through the soil limit moisture available for plant growth. Rain can easily dislodged and displaced soil particles.
- Reduce percolation makes soils difficult to wet and difficult to dry.
- High densities: Soil can become hard as concrete.
- Extreme temperature variations along vertical and horizontal gradients.

- ➢ Increase physical barrier (concrete, asphalt, oil, etc.) with particle packed more tightly, roots have more difficulties in penetrating soil.
- ➢ Higher alkalinity from lime applications and breakdown on man made materials, imbalance of nutrients results with calcium predominating.

Increase metal content from car exhaust and industrial emission, exposure to road salt destroy soil structure upsets ionic balance. Trees with limb susceptible to breakage under moderate wind can be dangerous as street trees. Trees with thick bark are vulnerable to mechanical damage by people and vehicles. Drought adapted trees with deep root system can anchor themselves in areas continuously exposed to high wind. Trees with stout rigid, or highly flexible trunk can both withstand high wind such as palm trees.

REFERENCES

"Catchment degradation in Asia and Pacific", (1984). Soil and Water, issue No. 3, New Zealaur.

Evan, J. W. (1981). *Preliminary Post Moretem of Watershed Management, Projects in Thailand* for National Environmental Board, Thailand.

Luding, H. F. (1986). *Manual of Environmental Guidelines for Watershed Development Project*, For ADB, Manila.

Eley, J. (1984). *Engineering Hydrology*, Civil & Environmental Engineering, Utah State University.

Mac Kinnon J., A *National Conservation Plan for Indonesia*, Vol. III, National Park development and General Topics, II FAO, United Nations, Bogor, Appendix 44-2/98.

Giles, R. H. (ed.) (1971). *Wildlife Management Techniques*. Brd. Ed., The Wildlife Society, Washington D.C.

Conant, I. *et al.* (eds.) (1983). *Resource Inventory & Base Line Study Methods for Developing Countries*, American Association of Advancement of Science, Washington D.C.

Manual of Environmental Technology in Developing Countries, (1991). Vol. II, Harvey Ludwig, Warran Evans, Warram Y. Brock LMAM, Bindu Lohari, South Asia Publishers Pvt. Limited, New Delhi, Unpublished.

Research Report, (2002). *Landscape Restoration Principles and Practices: An Indian Context*, Arup K. Dewan.

Unpublished Research Report, (2002). *Issues Associated with Management of Indian Forest: An Approach*.

10

Influence of Soil Degradation on Soil Properties in Hilly Tropical Forest Soils of Arunachal Pradesh

S. C. TIWARI and S. S. SINGH

Introduction

Soil degradation is the manifestation of undesirable changes of physico-chemical and biological characteristics occurred in the soil environment as a result of disturbance. Deforestation for expansion of shifting cultivation (jhum) and selective logging is the cause of reduced fertility and degradation in the hilly forest soils of humid tropical regions of North Eastern India. Conversion of forests to intensive agriculture often leads to degradation of weathered soil (Caldwel, 1999), and is expected to produce, in the long term, a decline in soil organic matter and soil fertility. The changes in microbial biomass 'C' following vegetation removal can provide an easy indication for slower and less easily detectable soil organic matter, hence fertility level of soils (Henrotanl Robertson, 1994). Microbial activities of the soil, despite their importance in many of the soil processes, are frequently disturbed, as shown by altered enzyme activities and microbial biomass, following agricultural exploitation and tillage practices. Monitoring soil physico-chemical properties may provide useful information in assessing the sustainability of agricultural systems as these parameters are responsible for improving soil fertility status and higher plant growth (Rew *et al.* 1997). Microbial activities in terms of microbial biomass, and enzyme activities are affected by deforestation, agriculture, tillage and soil management practices. The measurement of these parameters can be used as sensitive indicators of ecological stability (Ajwal *et al.* 1999), soil quality (Kennedy *et al.*, 1995), total microbiological activity and fertility level of soils (Dkhar, Tiwari, and Steinberger, 1983; 1988; Chander, 1981; Henrot, 1994).

Therefore, the present study is an attempt to evaluate the influence of degree of soil degradation on microbial biomass carbon, dehydrogenase activity and other soil physico-chemical properties in hilly forest soils experiencing varying degree of forest degradation.

Material and Methods

The Banderdewa forest range was selected as an experimental site for the present study. The study site is located between 27° 6′ N latitude and 93° 49′E longitude at an elevation of 350 m amsl. The soil in the study site falls under Karsingsa series, which is a member of sandy mixed hyperthermic family of Typic Udifluvents. The mean annual precipitation in the last three years ranged from 3400 to 5700 mm of which 80 per cent is received during summer rainy season from March to August and 20 per cent during the winter dry period from November to February. The mean ambient temperature is 26°C. The natural vegetation consists of semi-evergreen and mixed deciduous forests. The presence of a protected forest area in the range allowed us to select an undegraded natural forest (UDF) as control site and two others, degraded forest (DF) and slightly degraded forest (SDF) sites for the present study. The selection of study sites was based on living tree stump density, species composition of dominant trees and land use history (Tables 10.1 and 10.2).

Degraded forest site (DF) is a regenerating jhum fallow land which had been used for 13 years for cultivation of rice, maize, finger millet and tuber crops. This site has not been used for further cultivation of crops since last jhum cultivation practiced in 1993 due to low crop productivity and currently lying as an abandoned fallow land. The soil surface is covered with luxuriant growth of fern and grasses during summer rainy season while dried and plant residue covers the soil surface in winter dry seasons. The tree density is only 21 trees 100 m^{-2}. The SDF and UDF sites have not been used for any type of cultivation in the past. However, at the SDF site, clearing and burning of weeds were practiced by forest personals, for a gap of every 2 to 3 years to allow proper growth of forest plantations after selective logging. The forest floor is covered with litter, weeds and grasses throughout the year. The tree densities in SDF and UDF are 62 and 87 trees 100 m^{-2} respectively. The tree species composition in UDF is almost same with SDF site however, *Pterispermum acerifolium*, *Chukrasia tabularis* and *Artocarpus chaplasa* are the dominant trees occurred only in this site. The official record of the range reveals that the site consists of natural trees only. The forest floor is covered with thick layer of litter, weed and grasses throughout the year.

Table 10.1: Distribution of Dominant Tree Species in Degraded (DF), Slightly Degraded (SDF) and Undegraded (UDF) Forest Soils

Tree Species	*DF*	*SDF*	*UDF*
Ailanthus grandis Prain			
Altinga excelsa Noron	–	+	+
Artocarpus chaplasha Roxb	–	–	+
Bombax ceiba Linn.	–	+	+
Canarium strictum Roxb.		–	+
Chukrasia tabularis Andr. Juss.	–	+	+
Dillenia indica Linn.	+	+	+
Duabanga grandiflora (Roxb. ex. D.C.) Walp	+	+	+
Gmelina arborea Roxb.	+	+	+
Kydia calycina Rob.	–	+	+
Manglieta caveana HK. f. & Th.	+	+	+
Phoebe goalparensis Hutch.	–	+	+
Pterispermum acerifolium Willd.	–	–	+
Terminalia myriocarpum Huerck and Muell Arg.	–	+	+
Toona ciliata Roem.	–	+	+

"+" Means species present; "–" species absent.

Table 10.2: Particle Size Distribution, Textural Class, Degree of Soil Disturbance, Slope and Aspect of Degraded (DF), Slightly Degraded (SDF) and Undegraded (UDF) Forest Soils

	DF	*SDF*	*UDF*
Sand (%)	72.8	73.0	74.9
Silt (%)	20.1	18.0	18.1
Clay (%)	7.1	9.0	7.0
Textural class	Loamy sand	Loamy sand	Loamy sand
*Degree of soil degradation (%)	80	40	00
Slope (%)	15	15	15
Aspect (direction of slope)	NEW	NEW	NEW

* Percentage of degree of soil disturbance was calculated based on number of living tree stumps in DF and SDF as compared to the UDF site.

The present study was conducted for six months from August 1998 to January 1999, at monthly intervals in the selected sites. Sampling was done from three replicate plots of 100 m^2 each for all the three sites at the middle of every month, at least 24 hour later if rain had occurred. Soil samples were collected separately from two depths; surface (0-20 cm) and sub-surface (20-40 cm) layer after discarding approximately 0.5 cm soil in between the two depths using metal core soil sampler having diameter and height of 5.5 and 20 cm respectively. A total of 15 sample cores were collected from each replicate plot of DF, SDF and UDF sites and mixed to obtain a composite sample for each replicate plot. Soil parameters requiring fresh samples were determined immediately otherwise stored below 4°C until further use. Fresh samples were used for the determination of bulk density (Okalebo *et al.*, 1993). Moisture content was determined by drying 10 g fresh soil at 105°C for 24 h. The soil *p*H was measured in 1 : 2.5 (soil: water) suspension using electronic digital *p*H metre (Systronics 335, Ahemdabad, India). Chloroform fumigation technique was used for the determination of organic and microbial biomass carbon. Soil dehydogenase activity was assayed using the modified TTC (2, 3, 5- tri phenyl tetrazolium chloride) reduction technique (Casida, 1977). The overall mean values of the parameters studied for all the sampling periods were expressed on oven dry weight basis.

The relationship among studied parameters was analyzed by calculating correlation coefficient values (r). The differences of studied parameters among the three sites were analyzed statistically using one way analysis of variance (ANOVA). Origin 4.1 software was used to process statistical analysis of data.

Results

The percentage sand, silt, and clay contents and textural classes of soils in all the three study sites are given in Table 10.2. The textural class of soils in all the three sites is loamy sand. The overall mean and standard deviation (SD) values of soil physico-chemical properties are given in Table 10.3. The maximum bulk density (1.3 and 1.4 g cm^{-3}) was recorded from DF and minimum values (1.1 and 1.2 g cm^{-3}) were recorded from UDF soil at 0-20 and 20-40 cm soil depths respectively. However, the SDF soil had intermediate value of BD at both soil depths as compared to DF and UDF soils. SMC did not vary much in all the three sites. However, soil moisture content was lower at 20-40 cm soil depths in all the three sites. The soil reaction was more acidic in DF as compared to the SDF and UDF sites (Table 10.3). The maximum C_{org} (3.1%) was recorded in UDF, which was followed by SDF (2.4%), and DF (2.0%) sites at 0-20 cm soil depth. The C_{org} was reduced by 40 per cent at 0-20 cm and 20 per cent at 20-40 cm in DF and SDF soils as compared to the UDF soil. The C_{org} content was comparatively higher at surface than in sub-surface soil layer.

Table10.3: Physico-chemical Properties of the Three Soil Ecosystems in Degraded (DF), Slightly degraded (SDF) and Undegraded (UDF) Forest Soils at 0-20 cm and 20-40 cm Soil Depths

Study sites	*Soil depth (cm)*	*BD (g-1cm)*	*SMC (%)*	*pH*	*C_{org} (%)*
DF	0-20	1.3 ± 0.10	18.0 ± 11.68	5.5 ± 0.38	2.0 ± 0.51
	20-40	1.4 ± 0.08	16.2 ± 8.23	5.2 ± 0.21	1.2 ± 0.42
SDF	0-20	1.2 ± 0.06	18.0 ± 9.96	5.7 ± 0.23	2.4 ± 0.41
	20-40	1.3 ± 0.07	15.0 ± 4.98	5.3 ± 0.21	1.5 ± 0.46
UDF	0-20	1.1 ± 0.07	19.3 ± 9.50	6.0 ± 0.45	3.1 ± 0.45
	20-40	1.2 ± 0.03	15.0 ± 5.99	5.7 ± 0.44	2.0 ± 0.44

Note: Values presented are means ± SD of six sampling months.

The mean values of C_{mic} contents in DF, SDF and UDF soils are depicted in Figure 10.1. The C_{mic} was maximum in UDF (259 mg C g^{-1} soil) followed by SDF (215 mg C g^{-1} soil) and minimum (137 mg C g^{-1} soil) in DF sites at 0-20 cm soil depth. The soil of DF had lowest C_{mic} content and it increased in a trend SDF > UDF. There was a 50 per cent reduction of C_{mic} in DF site and 20 per cent in SDF soils as compared to UDF, which had the maximum C_{mic} at both soil depths. The sub-surface soil layer contained lower C_{mic} as compared to the surface soil layer.

The DHA in DF, SDF and UDF site for 0-20 cm and 20-40 cm soil depths are depicted in Figure 10.2. The maximum DHA (112 mg TPF g^{-1} soil 24 h^{-1}) was recorded from UDF which declined to 68 mg TPF g^{-1} soil 24 h^{-1} in SDF and DF sites respectively at 0-20 cm soil depth. The DHA was decreased by 40 per cent at 0-20 cm and 25 per cent at 20-40 cm soil depths in DF and SDF respectively, as compared to UDF soil. The surface soil contained higher DHA than sub-surface soil layer. The calculated values of correlation co-efficient (r) among physico-chemical properties, microbial biomass and dehydrogenase activity in DF, SDF and UDF are given in Table 10.2.

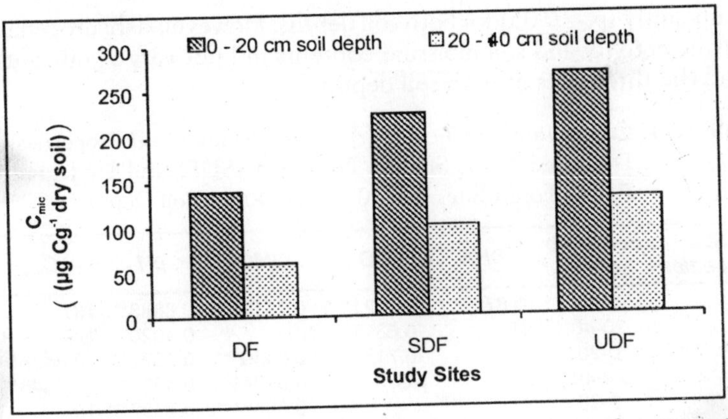

Figure 10.1: Variation in Microbial Biomass Carbon (C_{mic}) Contents in Degraded (DF), Slightly Degraded (SDF) and Undegraded (UDF) Forest Soils

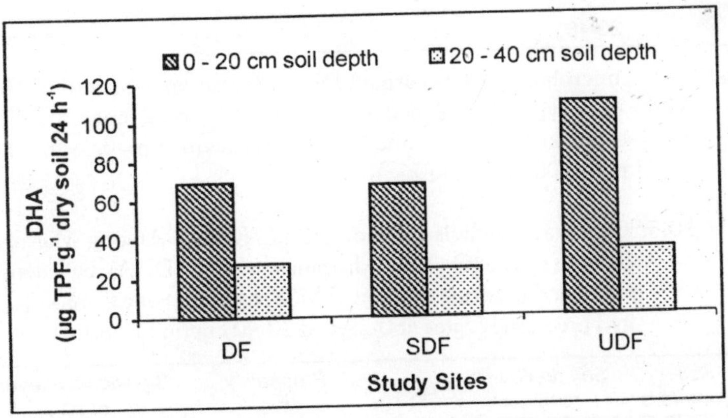

Figure 10.2: Dehydrogenase Activity (DHA) in Degraded (DF), Slightly Degraded (SDF) and Undegraded (UDF) Forest Soils

Microbial biomass carbon had significant (p = 0.05) positive correlation with DHA, SMC, and *p*H. However, microbial biomass carbon was negatively correlated to bulk density in all three sites at both the soil depths. Dehydrogenase activity was also correlated positively with SMC, *p*H and total organic carbon at both depths except in 0-20 cm soil depth where SMC showed negative correlation with DHA and BD. The results of one-way analysis of variance (ANOVA) of studied parameters are depicted in table 10.5. Microbial biomass carbon, total organic carbon, *p*H and bulk density varied in significantly (p = 0.05) for both soil depths. However, dehydrogenase enzyme activity and soil moisture contents did not vary significantly in all the three sites at both soil depths.

Table 10.4: Correlation Coefficient (r) Values Among Soil Properties in Degraded (DF), Slightly Degraded (SDF) and Undegraded (UDF) Forest Sites at 0-20 and 20-40 cm Soil Depths

Soil properties	*Soil depth (cm)*	*DHA*	*BD*	*SMC*	*pH*	*C_{org}*
C_{mic}	0-20	0.8797***	-0.7921***	0.7717***	0.8883***	NS
	20-40	NS	-0.6311**	NS	0.4020*	NS
DHA	0-20		-0.7185***	0.6440**	0.9072***	0.4425*
	20-40		NS	-0.6138**	0.4935*	0.6643***
BD	0-20			-0.6646***	-0.8657***	NS
	20-40			-0.4185*	NS	NS
SMC	0-20				0.7547***	-0.4661*
	20-40				-0.5369**	-0.6497**
pH	0-20					0.3992*
	20-40					0.6165**

Note: C_{mic}, microbial biomass carbon; DHA, dehydrogenase activity; BD, bulk density; SMC, soil moisture content; C_{org}, total organic carbon. Values marked with *, ** and *** are significant at p=0.05 , p=0.01 and p = 0.001 levels respectively.

Table 10.5: One way Analysis of Variance (ANOVA) Among Microbial Biomass Carbo (C_{mic}), dehydrogenase activity (DHA), bulk density (BD), soil moisture content (SMC), total Organic Carbon (C_{org}) in Three Study Sites at 0-20 and 20-40 cm Soil Depths

Soil parameters	*Soil depth (cm)*	*F ratio*	*Probability level*	*Significance level of F**
C_{mic}	0-20	4.15	p=0.05	Significant
	20-40	6.65	p=0.05	Significant
DHA	0-20	3.31	p=0.05	Not significant
	20-40	1.79	p=0.05	Not significant
BD	0-20	9.73	p=0.05	Significant
	20-40	14.47	p=0.05	Significant
SMC	0-20	0.04	p=0.05	Not significant
	20-40	0.81	p=0.05	Not significant
pH	0-20	2.11	p=0.05	Significant
	20-40	5.31	p=0.05	Significant
C_{org}	0-20	10.19	p=0.05	Significant
	20-40	3.59	p=0.05	Significant

* Degree of freedom 11.

Discussion

The microbial biomass carbon and dehydrogenase activity declined after forest soil disturbance by jhum cultivation and selective logging practices. Henrot and Robertson (1994) reported decline in microbial biomass carbon following vegetation removal in Costa Rican humid tropics. Microbial biomass carbon was comparatively higher in the UDF site than DF and SDF sites. Hajabassi *et al.* (1997) also reported 20 per cent increase in bulk density, 50 per cent decrease in organic matter, lowered soil quality and ultimate decline in productivity of natural soils due to deforestation in Iraq. This is in agreement with the present findings of 10-20 per cent increase of soil bulk density and decreased organic and microbial biomass carbon contents in SDF and DF sites following forest degradation due to jhum cultivation and selective logging practices. The activity of DHA also decreased by a magnitude of 40 per cent at surface soil and 25 per cent in sub-surface soil layer for degraded and slightly degraded soils as compared to the undegraded natural forest soil. Beyer *et al.* (1999) had reported decreased microbial biomass C and dehydrogenase activity of soils with a low total organic carbon level compared to soils with a typical total organic carbon level which is in correspondence with the findings of present study in which DF and SDF soils had comparatively lower total organic carbon, microbial biomass carbon and dehydrogenase activity at surface and sub-surface soil layers as compared to the UDF soil. The decreased microbial biomass carbon content and low dehydrogenase activity had also been reported (Batra and Manna, 1997) leading to lower biological activities (Garcia, 1997 from degraded soils. The highest microbial biomass carbon content and dehydrogenase activity in UDF site indicates that large microbial biomass accumulation increased their activity in presence of higher organic matter, favourable soil reaction and soil moisture content (Dhillon *et al.*, 1996). The positive correlation between the dehydrogenase activity with moisture content, *p*H and organic carbon contents in this study is supported by the reports of Henrot and Robertson (1994), Perucci (1997) and Dinesh (1998) which indicates that these physico-chemical properties play important roles in detecting the levels of soil microbial biomass and enzyme accumulation in degraded and undegraded soil ecosystems in revealing the fertility status of the soil ecosystems in terms of microbiological and biochemical activities.

Conclusion

The present study demonstrates that soil degradation due to jhum cultivation and selective logging caused greater reduction in physico-chemical soil properties, namely bulk density, microbial biomass carbon levels and dehydrogenase activity.

Acknowledgements

Authors are thankful to CSIR, New Delhi for financial assistance and Director, NERIST for facilities.

REFERENCES

Ajwa, H. A., Dell, C. J. and Rice, C. W. (1999). Changes in enzyme activities and microbial biomass of tall grass prairie soil as related to burning and nitrogen fertilization. *Soil Biology and Biochemistry* 31: 769-777.

Batra, L. and Manna, M. C. (1997). Dehydrogenase activity and microbial biomass C in salt-affected soils of semi-arid and arid regions. *Arid Soil Research and Rehabilitation* 11: 293-303.

Beyer, L., Seiling, K. and Pingpank, K. (1999). The impact of a low humus level in arable soils on microbial properties, soil organic matter and crop yield. *Biology and Fertility of Soils* 28:155-161.

Caldwel, B. A., Griffiths, R. P. and Sollins, P. (1999). Soil enzyme response to vegetation disturbance in two lowland Costa Rican soils. *Soil Biology and Biochemistry* 31: 1603-1608.

Casida, L. E. (1977). Microbial metabolic activity in soil as measured by dehydrogenase determinations. *Applied Environmental Microbiology* 34: 630-636.

Chander, K. and Brookes, P. C. (1991). Is dehydrogenase assay invalid as a method to estimate microbial activity in copper contaminated soil? *Soil Biology and Biochemistry* 23: 630-636.

Dinesh, R., Dubey, R. P. and Prasad, G. S. (1998). Soil microbial biomass and enzyme activities as influenced by organic manure incorporation into soils of rice-rice system. *Agronomy Crop Science* 181: 173-178.

Dhillon, S. S., Roy, J. and Abrams, M. (1996). Assessing impact of elevated CO_2 on soil microbial activity in a Mediterranean model ecosystem. *Plant Soil* 187: 333-342.

Dkhar, M. S. and Mishra, R. R. (1983). Dehydrogenase and Urease activities of maize (*Zea mays* L.) field soils. *Plant Soil* 70: 327-333.

Garcia, C. and Hernandez, T. (1997). Biological and biochemical indicators in derlict soils subject to erosion. *Soil Biology and Biochemistry* 29: 171-177.

Hajabassi, M. A., Jalalian, A. and Karimjadeh, H. R. (1997). Deforestation effect on soil physical and chemical properties, Lordegan, Iran. *Plant Soil* 190: 301-308.

Henrot, J. and Robertson, G. P. (1994). Vegetation removal in two soils of humid tropics: Effect on microbial biomass. *Soil Biology and Biochemistry* 25: 111-116.

Kennedy, A. C. and Papendick, R. I. (1995). Microbial characteristics of soil quality. *Journal of Soil Water Conservation* 50: 243-248.

Okalebo, J. R., Gathua, J. W. and Woomer, P. L. (1993). *Laboratory Methods of Soil and Plant Analysis: A Working Manual.* TSBF-Soil Sci. Soc. East Africa Tech. Publ. No. 11993.

Perucci, P., Bonciarelli, U., Santilocchi, R. and Bianchi, A. A. (1997). Effect of crop rotation, nitrogen fertilizer and management of crop residues on some chemical, microbiological and biochemical properties of soil. *Biology and Fertility of Soils* 24: 311-316.

Rao, A. V., Singh, K. C. and Gupta, J. P. (1997). Ley farming, An alternative farming, system for sustainability in the Indian arid zone. *Arid Soil Research and Rehabliitation* 11: 201-210.

Steinberger, Y., Vishnevetsky, S., Barness, G. and Lavee, H. (1998). Effect of topo-climatic gradient on soil dehydrogenase activity in a Judean desert ecosystem. *Arid Soil Research and Rehabilitation* 12:387-393.

Tiwari, S. C., Tiwari, B. K. and Mishra, R. R. (1988). Enzyme activities in soil: Effect of leaching, ignition, autoclaving and fumigation. *Soil Biology and Biochemistry* 20 : 583-585.

11

The Role of Geomorphic Impact Assessment in Land Utilisation Studies

N. C. JANA

SECTION I: INTRODUCTION, OBJECTIVE, DATA BASE, METHODOLOGY AND STUDY AREA

Introduction

Landscape, a combined effect of structure, process and stage greatly determines the degree of land use practices. The physical properties such as geology, nature of relief and underground water, drainage conditions, climatic phenomena, soil character, and vegetal cover etc. and their impact of land utilization are dealt with in applied geomorphology.

In the present juncture of development-environment nexus, the application of geomorphological knowledge in resource appraisal especially the land resources is more relevant to meet the increasing needs and aspirations of the human society. Thus, applied geomorphology means the application of geomorphological methods of survey and analysis in solving the physical environmental problems arising from flood control, land reclamation, etc. (Clark, 1985). Geomorphology, the science of landform, obviously can play a pertinent role in investigation, evaluation and classification of land and soil resources. Modern application of geomorophology is closely concerned with the interrelations among rock type, landform, drainage pattern, soil character and vegetation type. The distribution and character of rocks, minerals, soils, vegetation etc. are largely determined by the landform characteristics. So, the application of geomorphological knowledge is indispensable in the evaluation, classification and utilization of land. In land utilization especially in agriculture, it is not only necessary to know the general character of the relief and the value of surface material but also the present morphogenetic processes that are actively modifying the landscape (Fairbridge, 1968).

Applied geomorphology is related mainly to the problem of environmental resource management. It consists principally of analyzing and monitoring landscape forming processes in relation to surface materials and evaluating landscape forming processes in relation to surface materials and evaluating landscape changes that may arise from human interference (Dayal, 1990). According to Brunsden (1985) "*In applied geomorphology the aims are to assist in the efficient discovery, assessment and wise management of the earth's finite resources, to prevent environmental deterioration and to avoid or prevent natural hazards*". Applied geomorphologists have got recognition by their worthwhile contribution through quantitative analysis of process-form relations at the micro-level, interpretation and analysis of aerial photo and satellite imagery, their conscious involvement in interdisciplinary research in land and project surveys and environmental management programmes. As a result, geologists, hydrologists, geographers, engineers, agricultural scientists, planners, policy decision makers are becoming increasingly aware of the importance of applied geomorphology. Rapid progress in the interpretation of air photographs and evaluation of remote sensing imagery has already provided a wealth of information on land characteristics and assessment (Dayal, 1990). Diverse applications of geomorphology include:

- Topographic and thematic mapping of natural resources,
- Environmental studies, especially the survey of natural hazards (earthquakes, volcanism, landslides, avalanches, land subsidence, floods and droughts),
- Rural development and planning with greater emphasis on land utilization, erosion control, conservation strategy and river basin development,
- Urbanization for urban extension, site selection of mining, and
- Assessment of communication net-work, river and coastal engineering etc. (Verstappen, 1983).

In applied geomorphological study especially in the field of land utilization the concept of landscape ecology provides a more scientific base. In landscape ecological study, land is considered as an important resource. From the utilitarian point of view, the various attributes of land (called land resources), viz. climate, parent material, relief, soil, water, vegetation etc. are mutually exclusive but genetically and ecologically correlated.

A landscape is a natural arrangement of interrelated tracts of land, which is very complicated in nature. Accordingly, the study of landscape ecology is very much relevant for land use planning which is concerned with the ecology of mankind as well as of plants and animals and offers long-term benefit. Thus, the study of landscape ecology may be defined as the study of land attributes as the objects and variables of ecosystem as well as the processes which relate these objects and variables with the consideration of some key variable to be controlled by human intelligence (Vink, 1975).

In such type of studies, soil, water and terrain conditions play key role in regard to land use planning. So, the overall knowledge of land, water and soil condition is indispensable for understanding the past and present land use as well as for the prediction of future land use purposes.

Land use is thus the result of interaction of physical/geomorphic and socio-economic factors in macro and micro-regional levels. In plains, plateaus and mountainous terrain the scale of land use greatly varies. Similarly macro-regional land features (i.e., basins, ridges, floodplains, valley flanks, scarps, pediments etc.) and micro-regional landforms (i.e., levees, alluvial fan and cones, monadnocks, drumlins, sand dunes, oasis etc.) greatly affect the type of land use. Geomorphology sets the foundation and background of land use whether it is glacial or karst or arid or coastal or volcanic region or whatever it may be.

The effect of terrain condition on land use may be considered as direct. Slope in one hand directly affects land use because it sets limitations on cultivation, while on the other it affects indirectly the climate, position of water table, development of soil and drainage etc. For a planned land use practice, the dynamic development of slope and relief must be taken into account. In different geomorphic units the degree of land use practices can be assessed.

The plain topography with a negligible slope ensures thick soil layer, large amount of precipitated moisture, greater soakage and enriched ground water reserve. It ensures continuity of cultivation as well as facilitates the use of modern agricultural implements. Because of low gradient such topography assures extensive irrigation. In the plain region consolidation of land holding is more easily feasible. The cultivation of land (i.e., tillage, spading, hoeing, weeding, sowing and harvesting) in the plains is easier as compared to plateaus and hills.

The plateaus represent extensive upland terrain characterized by steep sidewalls, flat and rough top surface. Development of zonal soils, greater continuity of cultivation, large terraced field, considerable proportion of unterraced land, much easier and more frequent contact between farm and field, easier communication system are some of the important characteristics of plateau terrain.

In mountainous terrain, the steepness of gradient largely affects the amount and intensity of solar radiation, texture, structure and depth of soil, availability of soil moisture, terracing, continuity of cultivation, field pattern, development of transport, possibilities of irrigation development, the yield rate of output etc. It may be mentioned that due to the steepness of slope in the mountains, soil and water are not sustainable which are responsible for lesser degree of land use practices. Land can intensively be practiced only with the adoption of suitable management practices (De and Jana, 1997).

The study area exhibits contrasting surface configuration leading to the spatial discrimination of land use and cropping patterns. In the present paper the physical properties, viz. geology, climate, soil, vegetation and morphometric attributes of relief, slope and drainage have been analyzed for the assessment of their impact on land use.

Objective of the Present Study

The land is considered as an important resource. So an assessment and improvement of the use of land are essential to any significant increase in food supply especially in the developing countries of the world. In most of the developing countries there is an urgent need to extend and improve the exploitation of land resources to achieve the maximum output of crops, livestock and timber as well as to ensure the productive capability of land. This objective can rapidly be attained by giving emphasis on the planning of development and on the implementation of the plans. The planning and implementation both are fruitfully done after the initial classification of land as inventory of natural resources based on geomorphological attributes and other physical properties which are directly related to their ecological set up.

Geomorphology as the science of earth forms has important applications in the investigation, evaluation and classification of land as a resource. In this context, it is concerned with the interrelation between landform, rock type, soil character, drainage character and nature of vegetal cover. That is why its area of study is related to that of geology, hydrology, pedology, ecology, agriculture and so on. In all such subjects especially in agricultural land use, the applications of geomorophological knowledge becomes an essential tool, because the nature of the ground surface is a prime indicator of the character and distribution of rock, minerals, soils, drainage, vegetation, agricultural land use etc. The geographers especially the geomorphologists are thus urgently required for the investigation and planning of land use in the developing countries where land improvements are essential to any remarkable development in agriculture.

The aim of the present work is to find out the impact of physical properties on land use especially in the field of agriculture taking into consideration a case study in South-Western Midnapore, West Bengal, India.

Data Base

The entire study is based on both primary and secondary data. For secondary information several government, semi-government, institutions, private organizations, libraries and related administrative machineries have been approached for necessary data for all practical purposes. The detail databases used in the present exercise are as follows:

- Extensive fieldwork has been conducted to collect primary information regarding terrain characteristics, micro-relief, slope character, hydrological attributes, soil character, natural vegetation, environmental hazards etc.
- Soil samples have been collected during fieldwork and analyzed subsequently in the laboratory.
- Depth of watertable for pre-monsoon, monsoon, and post-monsoon periods has been measured directly in the field by the author.
- Survey of India topographical sheets have provided an excellent base to understand the realistic geographical situation throughout the entire study.

- The Geological Survey of India has provided maps and literatures related to geology.
- Data regarding forests and other related aspects have been collected from the Divisional Forest Office, Midnapore West.
- District Soil Conservation Office, Midnapore (West) has provided a lot of information regarding physical aspects of the study area.
- Meteorological data have been collected from the Indian Meteorological Department, Calcutta.
- Hydrological information have been consulted from the Central Ground Water Board (Eastern Region) and Office of the Subarnarekha Barrage Project, Midnapore.
- Irrigation and Waterways Directorate, Midnapore West Division, has provided maps and information concerning sources and means of irrigation.
- Data regarding land use particularly agricultural land use has been collected from the field as well as from the Principal Agricultural Office (Midnapore West Division), Block Development Office and Agricultural Extension Office of the concerned blocks, Bureau of Applied Economics and Statistics.
- The land use maps of Eastern India and District Planning Map (Midnapore), published by the National Atlas and Thematic Mapping Organisation, Department of Science and Technology, Government of India have been consulted.
- Daily newspapers and periodicals have also been followed for latest information relevant to land use studies.
- Data concerning socio-economic parameters have been collected through selected questionnaires.
- Remote sensing data have been used for Micro-regional studies.

Methodology

The entire exercise has been divided into three distinct phases, viz. pre-field methods, field methods and post-field methods.

Pre-Field Methods

- Collection and consultation of available maps such as topographical sheets, geological maps, soil maps, thematic maps and aerial photographs etc.

- Collection of available literatures and statistical information (both published and unpublished) related to geology, geomorphology, hydrology, climate, soil, vegetation, land use, irrigation, production etc. from secondary sources and their compilation, processing and interpretation for arriving at an understanding about the existing situation of the area concerned.
- Preparation of the base map.
- Chalking out of detail programme including site selection for in-depth field study, whether macro or micro level and plotting the sites on the base map before proceeding for field investigation.

Field Methods

After preparing the base map and having some knowledge about the area through available data and literatures the systematic field methods followed are given below:

- Morphological divisions and their associated land units are identified on the basis of topographical maps, so that survey work can be conducted in a systematic way.
- Block diagrams or sketches are drawn for each of the morphological divisions during fieldwork.
- Measurement of the general gradient of the area concerned by Abney level/Dumpy level that provides information about the drainage direction.
- Observation of the drainage pattern in the study area.
- Study of the underground water situation by measuring depth of wells in pre-monsoon, monsoon and post-monsoon periods in the area concerned.
- Collection of soil samples from the pre-selected sample sites, soil profiles in different pedo-geomorphic zones are also studied.
- Collection of data regarding the sources and means of irrigations as well as infrastructural and institutional facilities (through quetionnaries and other secondary sources).
- Identification of the general land use pattern which includes different types of cropland, culturable wastes, forest, horticulture, orchards, plantation sites, grazing land, fallow land etc.

- Study of the cropping pattern (mono, double and multiple cropping).
- Observation of change in land use pattern in each morphological region during the last years or so.
- Collection of production data directly from the farmers as well as from published information as and when necessary.
- Detection of problems in the field through personal observations as well as through questionnaires that are of great help in formulating planning strategies.
- The above data (physical, cultural etc.) for each of the morphological divisions are noted in tabular forms.
- Two micro level studies (using aerial photographs as available) have been taken to study the impact of physical properties on land use following the same procedure as above.
- Photographs of the typical features (both physical and cultural) in different seasons of the year are taken during the fieldwork.

Post-Field Methods

In post-field methods, data collected so far are compiled and analyzed. The systematic post-field methods followed here include:

- Finalization of the base map.
- Preparation and preservation of soil samples for subsequent laboratory analysis.
- Determination of *p*H, N.P.K., organic matter, soil moisture, texture etc. in the laboratory with the help of standard analytical method.
- Finalization of the schematic block diagrams (both macro and micro levels).
- Morphometric analysis of the study area based on S.O.I. topographical sheets to find out the morphological attributes (absolute relief, average relief, relative relief, drainage frequency, drainage density, dissection index, ruggedness index, slope etc.).
- Analysis of aerial photographs and mapping in the micro level studies.
- Finalization of the tabular descriptions of geomorphic attributes as well as present and prospective land uses.

- Evaluation of land potentialities of the morphological divisions based on available information.
- Final classification of the area under study based on capability and limitations.
- Finalization of the general land use (finally identified) including agricultural land use in the area.
- Correlation and analysis of physical properties and land use of each of the categories of land so far identified.
- Planning strategies are suggested accordingly for balanced regional development.

Study Area as a Whole

The South-western part of Midnapore District, West Bengal, India extends latitudinally from 21°44′ N to 22°26′ N and longitudinally from 86°42′ E to 87°33′ E. Comprising an area about 2392.88 sq. km (239288 hectares) it includes Gopiballavpur (I&II), Nayagram, Sankrail, Keshiary, Narayangarh, Dantan (I&II) Blocks (Figure 11.1).

As a whole the study area represents dual character of Chotanagpur plateau fringe in the west and Gangetic alluvial plan in the east. The geological formation of the area belongs to alluvium. Majority of the area shows beds of gravels, grits, and sands of Tertiary age. The surface mostly is of pisolitic, gravelly and nodular in character, while the eastern part as well as the river valley are alluvial in nature.

The topography here varies from moderately to gently undulating (north-western part) to almost flat plain (south-eastern part). The western and north-western boundary of the study area is more broken and picturesque in appearance. The general slope is from north-west to south and south-east.

The area is traversed by River Subarnarekha (golden-streaked river) and its tributary Dulung. Another important river is River Keleghai. The banks of the River Subarnarekha are generally high and well-defined and the cross profile is asymmetrical with a steeper slope towards the south-east. In summer, the River Subarnarekha dwindles away into a narrow and shallow stream while in rainy season flood is an occasional phenomenon.

The climate appears to be semi-continental. As a whole it has modified tropical monsoon climate. High temperature and seasonal rainfall with cyclonic disturbances are the conspicuous features. The mean January temperature is 20°C whereas the month of May records

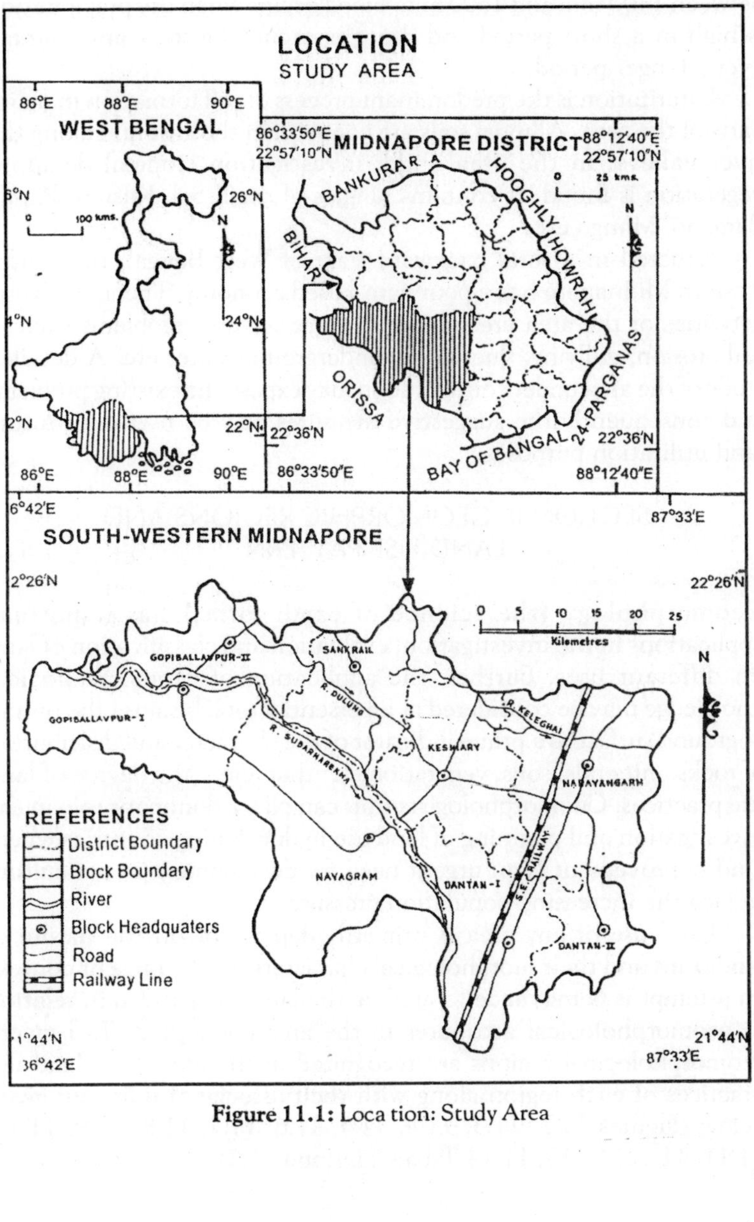

Figure 11.1: Loca tion: Study Area

32.35°C. Rainfall is erratic and irregular. Average annual rainfall varies between 1400 mm and 1600 mm. Flood occurs when the precipitation is high in a short period and drought occurs due to scanty rainfall over a longer period.

Laterization is the predominant process of soil formation in major parts of the area. Alluvial soils are common in the east and along the river valleys. In the area under investigation tropical decidous vegetation is found. It contains jungles of small Sal, Kusum, Piasal, Bamboo, Mango etc.

Situated in a rural backward tract of West Bengal, the south-western Midnapore has a poor agro-based economy. The agricultural activities of the area are subject to some adverse problems such as soil erosion, gullying, fluctuating underground water etc. A detailed study of the area under consideration may expose the existing problems and consequently the suggestive measures may be useful for future land utilization purposes.

SECTION II: GEOMORPHIC REGIONS AND LAND USE PATTERN

Geomorphology (the science of earth forms) has important applications in the investigation, evaluation and classification of land for different uses. Further, the application of geomorphological knowledge may be considered as an essential tool, because the nature of ground surface is a prime indicator of the character and distribution of rocks, minerals, soils, vegetation etc. that affect the degree of land use practices. Geomorphologists thus can play a dominant role in the investigation and planning of land use in developing countries where land improvement is an urgent need for development in agriculture to face the increasing population pressure.

Land use of any area is primarily dependent on the nature of landforms and their morphological characters. In the present context an attempt is being made to analyze the land use pattern in relation to geomorphological attributes in the area concerned. Two macro geomorphological regions are recognized in the area. The land use practices of each region along with their associated units are given below (Figures 11.2, 11.3, 11.4, 11.5, 11.6, 11.7, 11.8, 11.9, 11.10, 11.11, 11.12, 11.13, 11.14 Tables 11.1 and 11.2):

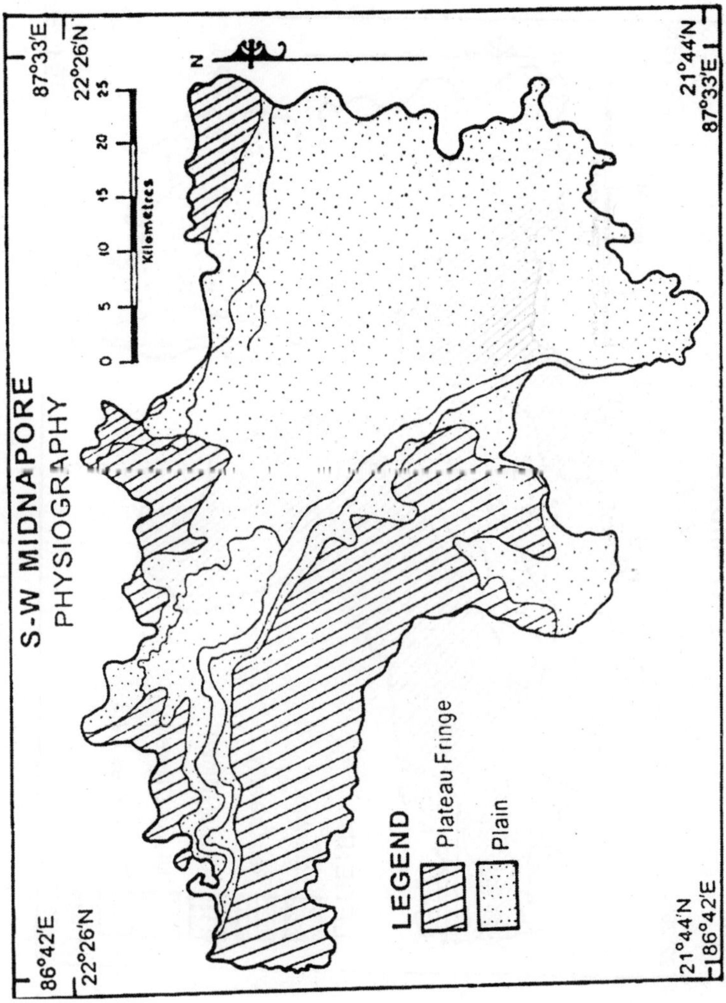

Figure 11.2: S-W Midrapore: Physiography

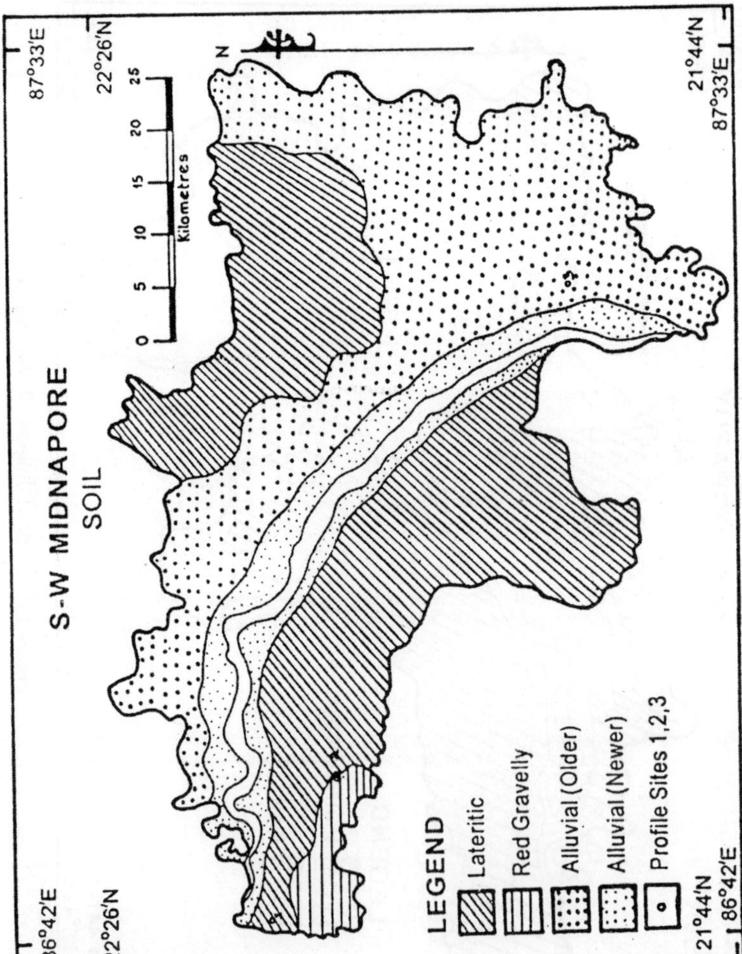

Figure 11.3: S-W Midnapore: Soil

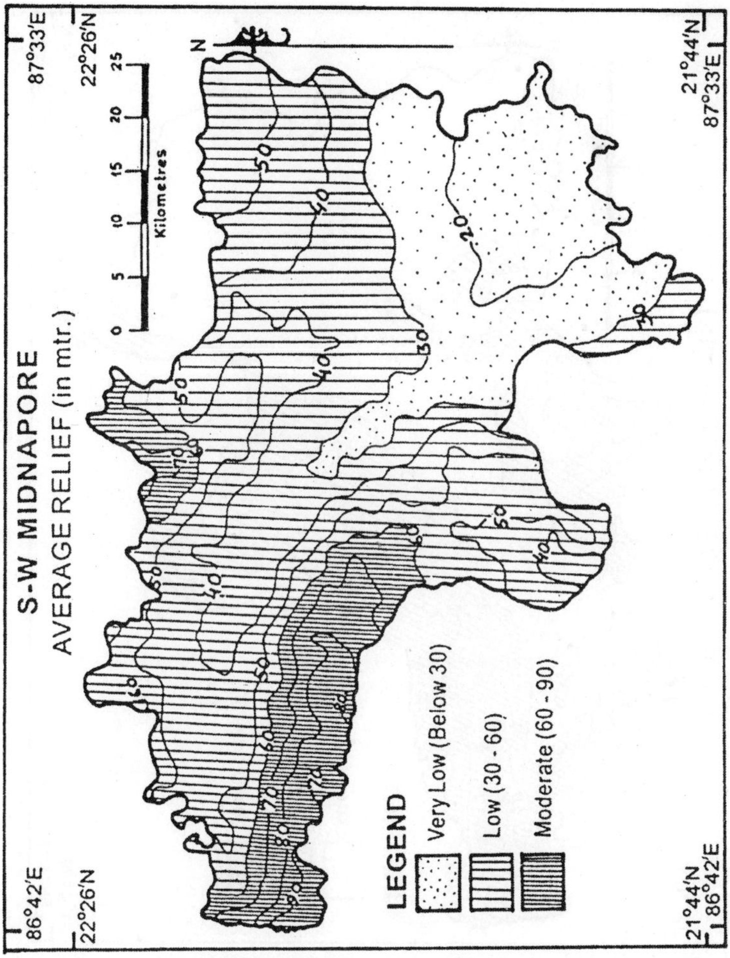

Figure 11.4: S-W Midnapore: Average Relief

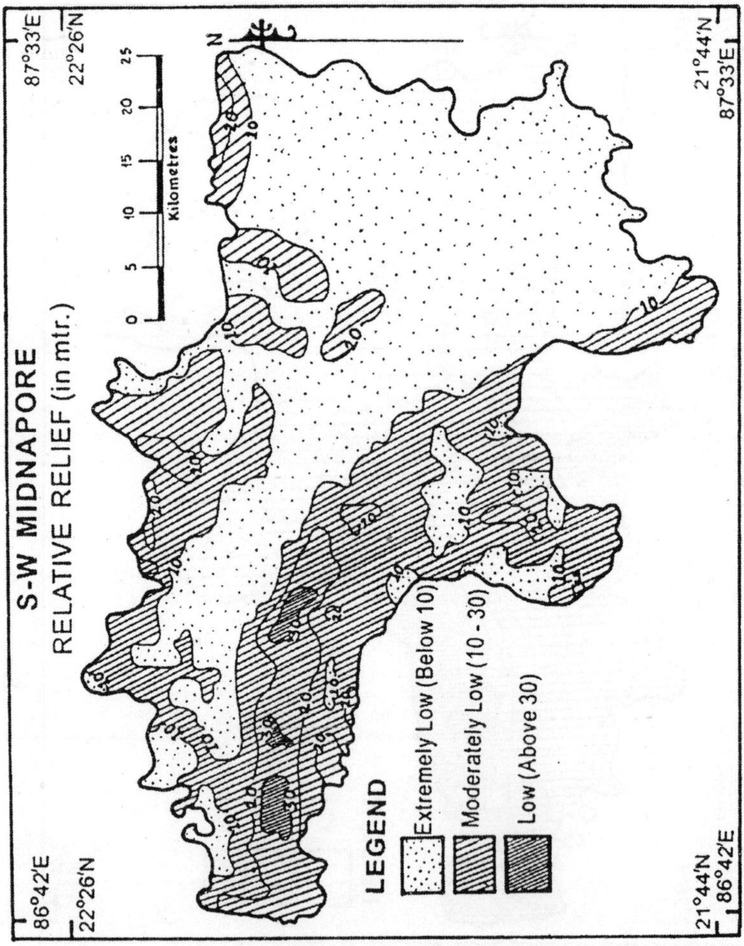

Figure 11.5: S-W Midnapore: Relative Relief

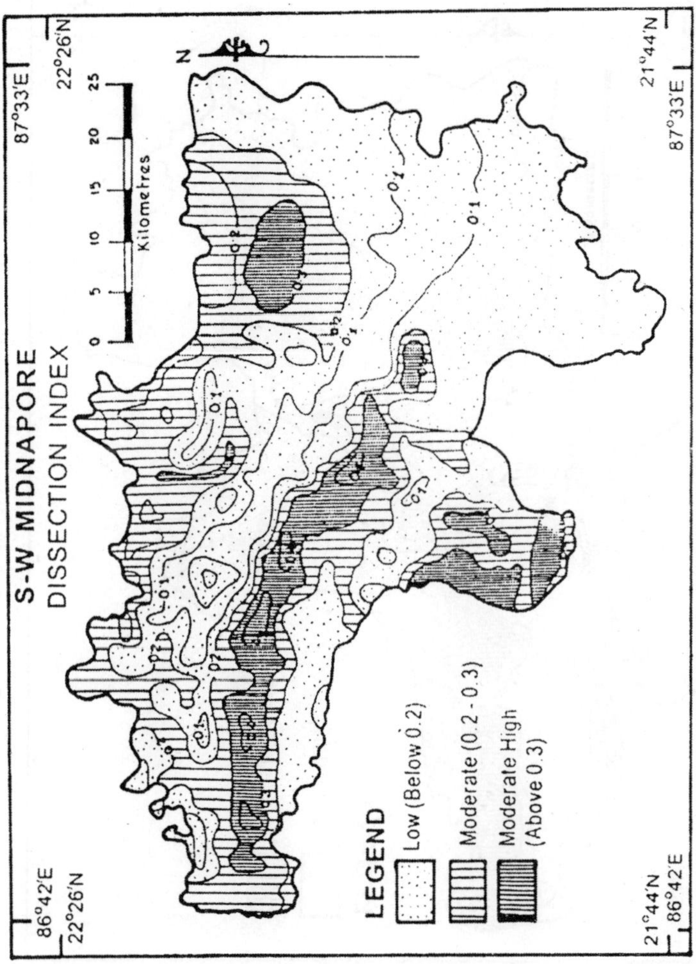

Figure 11.6: S-W Midnapore: Dissection Index

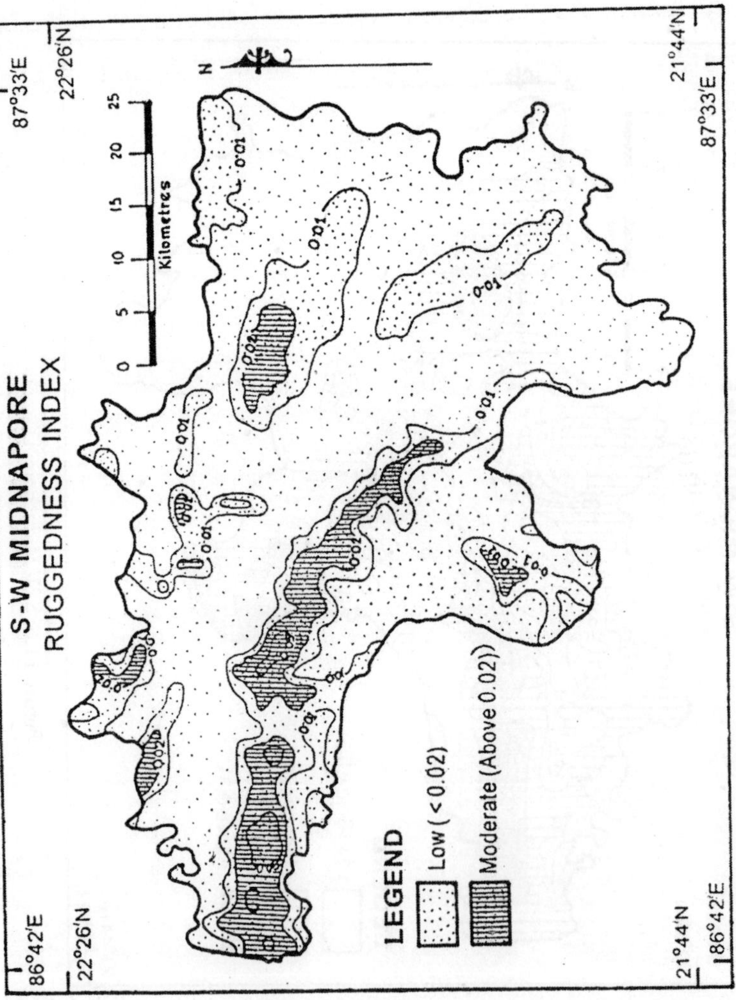

Figure 11.7: S-W Midnapore: Ruggedness Index

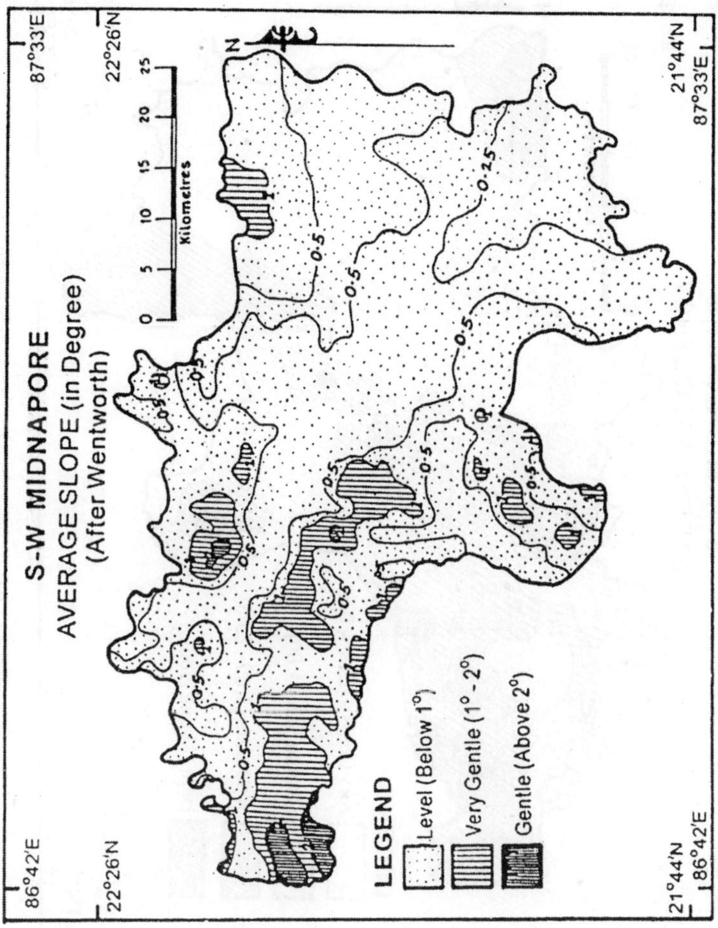

Figure 11.8: S-W Midnapore: Average Slope (in Degree) (After Wentworth)

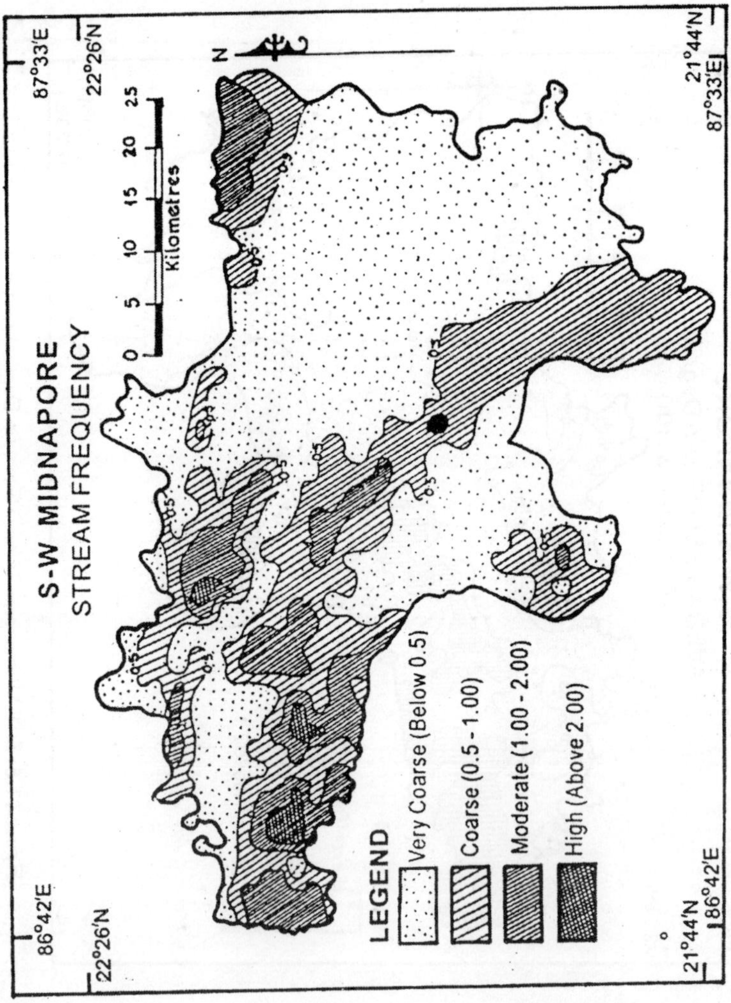

Figure 11.9: S-W Midnapore: Stream Frequency

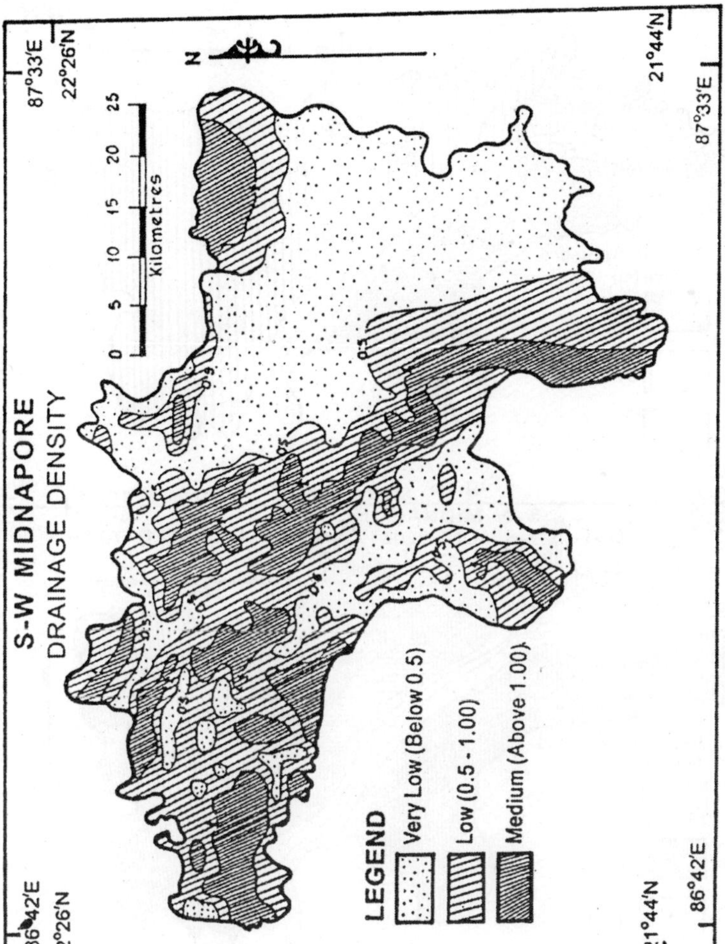

Figure 11.10: S-W-Midnapore: Drainage Density

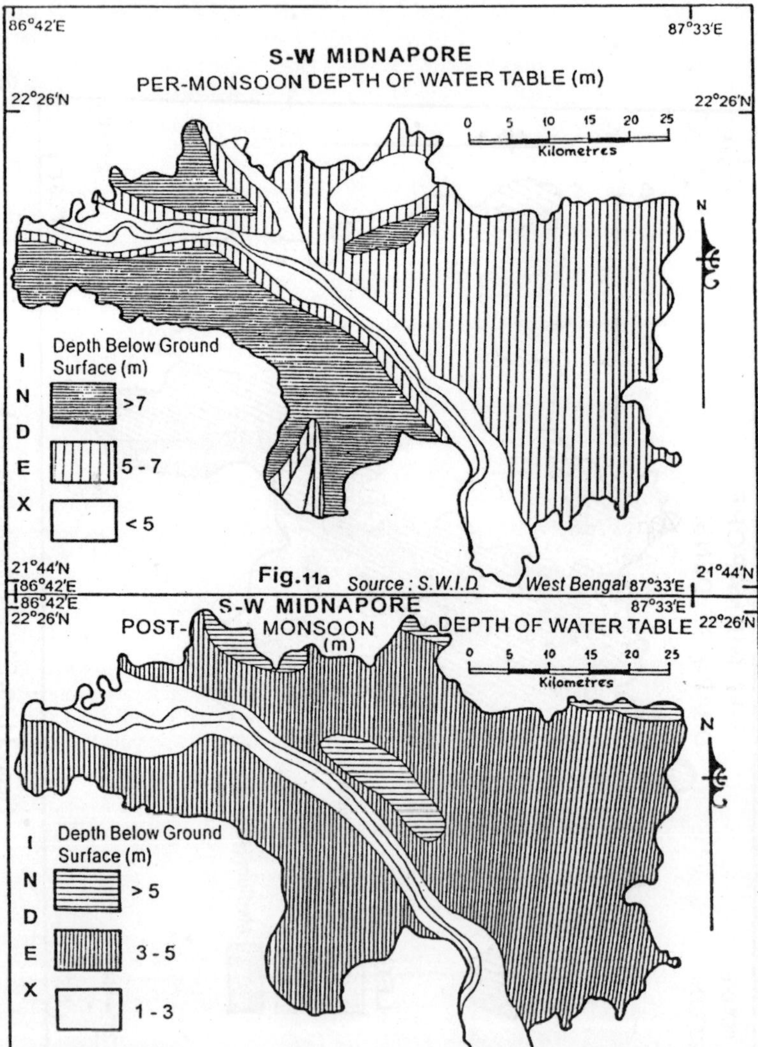

Figure 11.11: S-W Midnapore: Per Monsoon Depth of Water Table

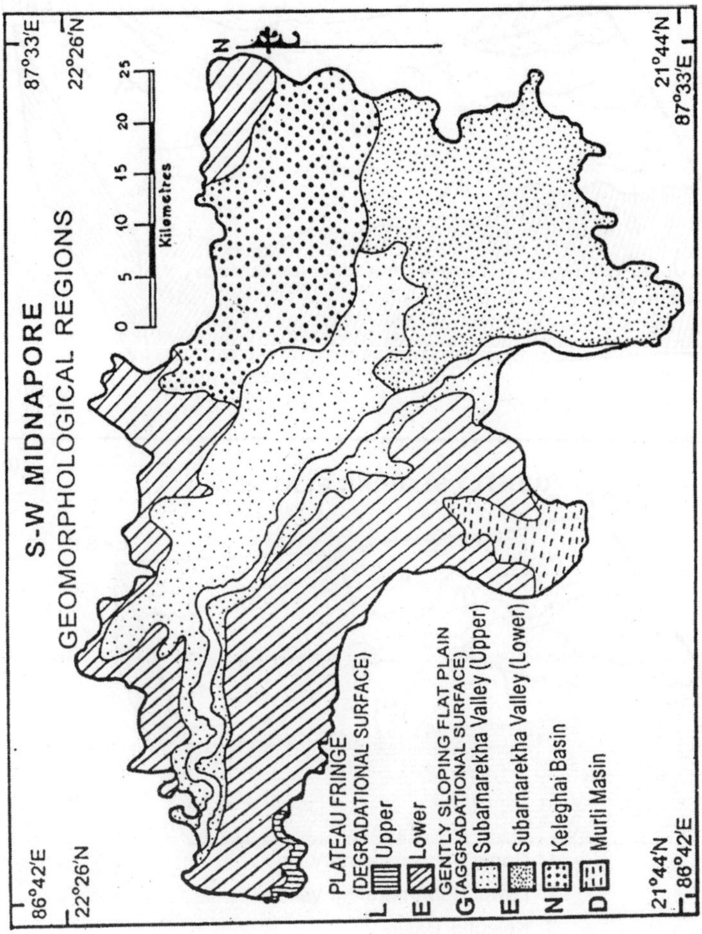

Figure 11.12: S-W Midnapore: Geomorphological Regions

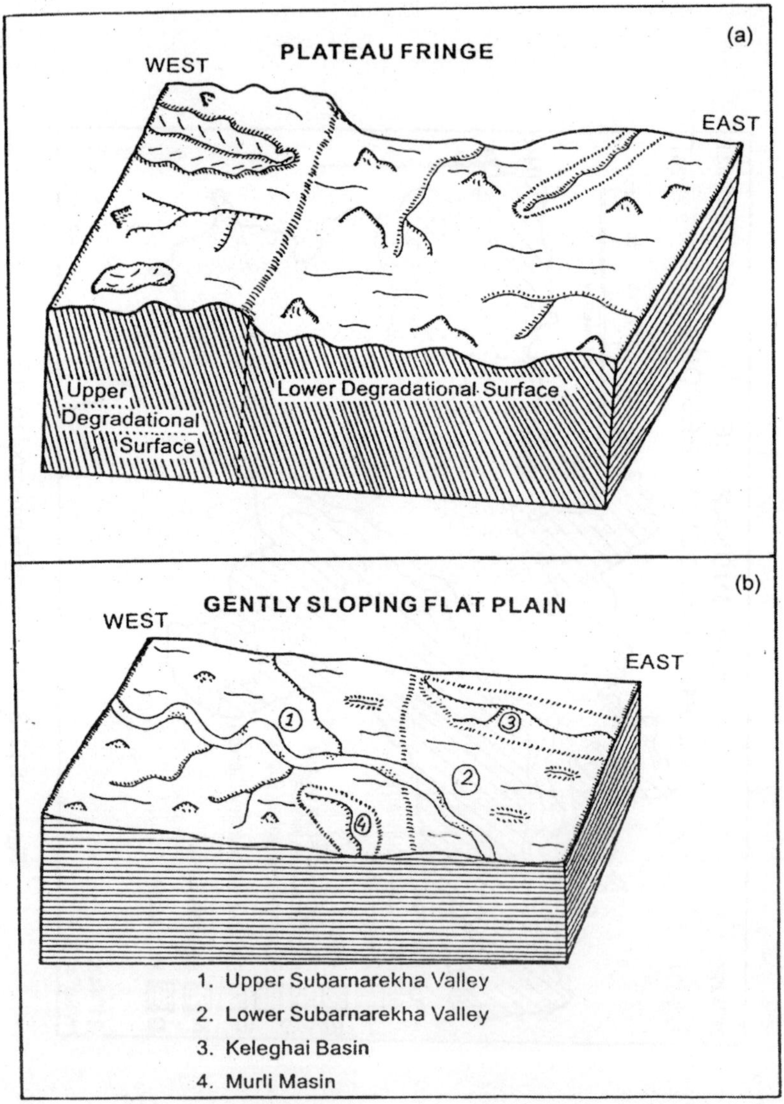

Figure 11.13: (a) Plateau Fringe and Gently Sloping Flat Plain (b)

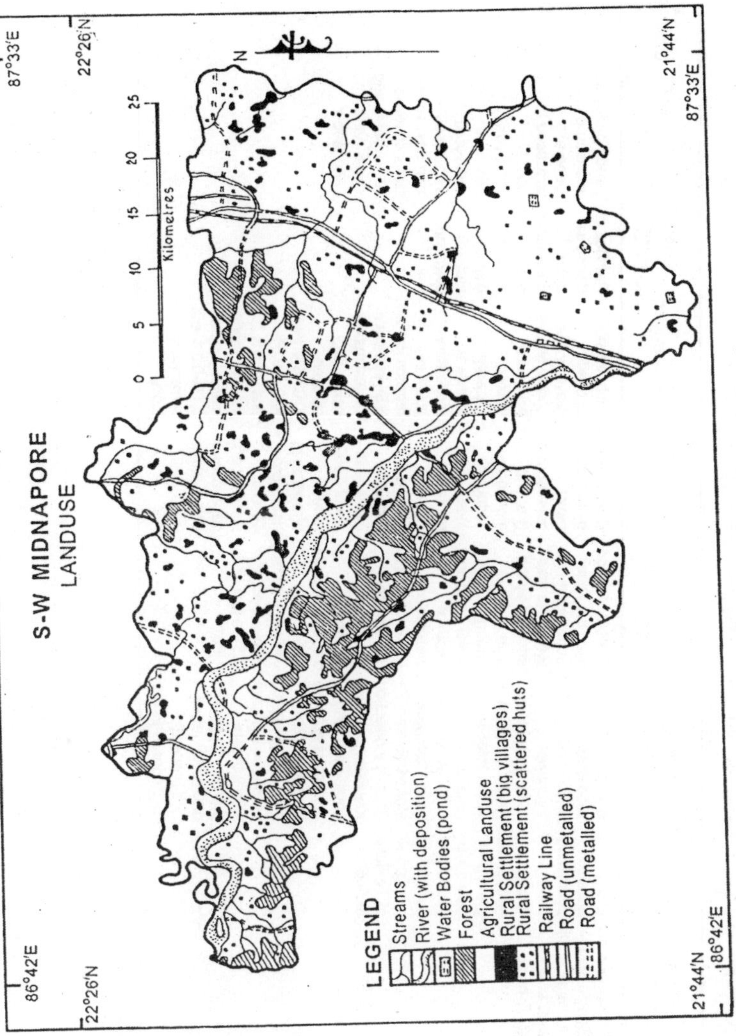

Figure 11.14: S-W Midnapore: Landuse

Table 11.1: Plateau Fringe (Degradational Surface)

Units	*Area Delimitation*	*Geological Foundation*	*Topographic Conditions*	*Hydrological Conditions*	*Climatic Phenomena*	*Soil Character*	*Vegetal Cover*	*Present Land Use*	*Land Quality/Potentiality*	*Land Class*
1. Upper Degradational surface	Small parts of Gopiballavpur-(extreme south west of the study area)	Older alluvium containing sand, laterities and lateritic gravels of Quaternary period and ferruginous sandstone of Tertiary period	Sharp to moderately undulating topography, average relief is moderate (60m-90m); relative relief is moderately low (30m); average slope between 2°-3°; slope direction: towards south and south-east; and surface is gravelly in nature with isolated uplands; erosional features with intricate gullies; moderate dissection index and moderate ruggedness index	Distant location from the River Subaranarekha; the unit is less affected by the river, poor in underground water condition; depth of waterable:> 7m (pre-monsoon) 3m-5m (post-monsoon); stream frequency is low (0.5-1.00) drainage density is low (below 0.5) to low (0.5-1.00)	Characterized by hot and moist conditions (tropical monsoon climate); monsoon rainfall in summer months (annual average is about 1500 mm), temperature rises upto 40°C in summer and decreases to 8°C in winter	Soils: lateritic and red gravelly, thin, coarse texture, low water holding capacity, slightly to moderately acidic in reaction (5.5 to 6.5); iron oxide context mainly below 3%; nitrogen (0.05%-0.1%); K_2O is 0.5% and P_2O_5 is 0.01%; organic matter content: 0.1%-0.3% and sometimes 0.5%	Patches of dense sal bush; eucalyptus, Mahua, Kusum; Bamboo etc. are common	Aman paddy as monocrop generally practiced in suitable locations; H.Y.V. is cultivated in a very small portions; some parts covered with forests and some parts lying fallow	P O O R	V

(Contd.)

Units	Area Delimitation	Geological Foundation	Topographic Conditions	Hydrological Conditions	Climatic Phenomena	Soil Character	Vegetal Cover	Present Land Use	Land Quality/Potentiality	Land Class
2. Lower Degardational surface	Ninety per cent of the plateau fringe; confined mainly to northern and north-western parts (Sankrail, Gopiballavpur-II and Nayagram), Southern and south-western parts (Nayagram and Gopiballavpur-I)	Older alluvium consists of beded lenticular layers of sandy and clayey deposits of Quaternary period; parts of the older alluvium may be of late Tertiary age	Moderately undulating topography, surface is gravelly, pisolitic and nodular in charcter, moderate average relief (60m-90m); low relative relief (above 30m); slope towards south and south-east and ranges between 1°-2° and above and some parts (depressions) below 1°; leaching predominant with gully erosion; soil erosion gets momentum by the torrential river during rainy season: and boulder extraction and deforestation; moderate to moderately high dissection index, low to moderate ruggedness index	Seasonal rivers of the chotonagpur intersect the lower degradational surface; River Subamarekha—principal river, Dulung-main tributary, drainage is comparatively by sharply cut banks; depth of water table: 5m-7m and above (premonsoon), 3m-5m (postmonsoon); sream frequency: medium to high (1.00-2.00 and above), very low to low (<0.5-1.00) in some parts; drainage density: low (0.5-1.00), very low (below 0.5) in some parts	Climate: semi-continental (modified tropical monsoon) high temperature and seasonal rainfall with cyclonic disturbances; mean January temperature is 20°C and temperature in May is 32.35°C; average annual rainfall: 1,400mm-1,600mm, suffers from hydrological drought	Mainly lateritic (northern and south-western) and older alluvial (north-western part); lateritic soils: sandy loam texture, reddish-brown colour, iron concretions present, completely bleached, low in nutrients; Fe_2O_3-below 5%, moderately acidic (*p*H: 5.5-6.0) N_2, K_2O, and P_2O_5 contents are 0.05%, 0.5% and 0.05%, organic matter content: 0.1%—0.5%; older alluvial soils (parts of the unit) are comparatively better than the lateritic soils in all respect	Lateritic tract dry deciduous type; luxuriant vegetation in the monsoon months with high rain-fall; completely leafless in hot dry weather, enriched with forest resources found in shallow soils and low soil moisture content, sal is predominant species; others include peasal, pial, Bhaira, Asan, Simul, Mahua etc., plantation species: Eucalyptus hybrid (main), Acacia, Cassia, Gammar etc.	Considerable part is devoted to forest, major forest products: timber, pole, fuel wood, wood pulp; minor forest products: sal seeds, Mahua seeds, Kend leaf, tassar etc. Aman as main crop generally practised in comparatively low lying areas where water is available; Babui grass (local name) cultivated in small parts of Nayagram and Gopiballavpur-1 Blocks	P O O R	V

1. Macro Regional Study

Plateau Fringe (Degradational Surface)

- *General*: This geomorphological region is an extension of the Chotanagpur plateau reflecting moderately undulating degradational surface.
- *Geology*: Mainly older alluvium (lateritic) consisting false bedded lenticular layer of sandy and clayey deposits with pebble beds.
- *Geomorphology*: The region is characterized by moderately undulating topography with slope towards south and southeast. The rivers originating from Chotanagpur intersect the region with sharply cut banks. The area is poor in underground water condition which becomes acute during summer season. Leaching is predominant in the area.
- *Climate*: Climate is characterized by hot and moist conditions (tropical monsoon climate). Average maximum temperature is 32.2°C in the month of May. Average annual rainfall is about 1500 mm. The region is invaded occasionally by cyclonic depressions.
- *Soil*: Soils are mainly lateritic having thin, coarse texture and low water holding capacity. Some parts are covered with red gravelly soil. These are relatively mature and are of poor quality.
- *Vegetation*: The region possesses flora closely similar to that of Chotanagpur. Sal is the predominant species. Others include Kusum, Asan, Mahua, Kend, Teak etc.
- *Land Use*: Rice is the main crop. Babui grass (local name) is cultivated in some parts. Significant part is devoted to forest cover (Table 11.1).

Gently Sloping Flat Plain (Aggradational Surface)

- *General*: The region occupies the eastern, south-eastern and the valley sections of the area under study.
- *Geology*: The geological formation is mainly newer and older alluvium. Large tracts of Holocene river-alluvium are found generally along the valley of the River Subarnarekha.
- *Geomorphology*: The region is an aggradational tract formed by the deposition of river-borne sediments. It is a gently

sloping flat terrain with low relief and level slope (towards east and south-east). River Subarnarekha and its tributary Dulung as well as River Keleghai and Murli Khal (canal) form the major drainage network. River Subarnarekha is characterized by wide braided channels reflecting conspicuous valley-side erosion. The rivers are sluggish in nature. As a whole the region experiences a very low intensity of soil erosion and almost free form flood ravages excepting some small pockets.

- *Climate*: Tropical monsoon climate (characterized by warm and moist conditions). Average maximum temperature is around 32°C recorded in the month of May. Average annual rainfall varies between 1500 mm and above 1600 mm. The region is affected by depressions and storms. In the post-monsoon period the region experiences a prolonged dryness.
- *Soil*: Mainly older and newer alluvial soils (along the valley). However, Murli basin and parts of the Keleghai experience lateritic soils.
- *Vegetation*: Tropical deciduous vegetation (Mango, Bamboo, Siris, Babla etc.)
- *Land Use*: Rice is the main crop. In addition, high yielding variety rice, wheat etc. are also grown. Sugarcane, potatoes, pulses and oilseeds are also practiced. The land use practices of each region along with their associated units are given in Table 11.2.

2. Micro Regional Study

Primarily based on the aerial photographs (1: 25,000) two micro-regional studies belonging to two macro geomorphological regions have been undertaken in the present context to establish relationship between geomorphic characteristics and land use pattern. Though the studies are based on aerial photographs, they have been supplemented by topographical maps and thorough ground check. The micro regions as identified in the photographs are:

- Keshiary-Bhasra Area
- Sorrong-Anikola Area

Table 11.2: Gently Sloping Flat Plain (Aggradational Surface)

Units	*Area Delimitation*	*Geological Foundation*	*Topographic Conditions*	*Hydrological Conditions*	*Climatic Phenomena*	*Soil Character*	*Vegetal Cover*	*Present Land Use*	*Land Quality/ Potentiality*	*Land Class*
1. Upper Subarna-rekha valley	Lies within Gopiballav-Pur-I & II, Nayagram, Sankrail and Keshiary Blocks	Newer alluvium (Quaternay period) consisting of sand with some silt and clay; general lithology is alluvial, sandy, and gravelly	Gently undulating plain with low relief although moderate in some parts; average relief: 30m-60m, Relative relief: extremely low (below 10m); slope: level (below 1°) to very gentle (1°-2°) land surface shows many depositional characteristics; river bank erosion common especially in the upper sections; depositional features with broad floodplain towards south-east; low dissection and ruggedness indices.	Subarnarekha is the only river, extreme upper parts (points of entrance) show sharply cut banks with rocky beds; wider flood-plain further south-east, stream frequency: very coarse (below 0.5) to coarse (0.5-1.0); drainage density: very low (0.5-1.00); depth of water table: <5m (pre-monsoon) and 1m-3m (post-monsoon)	Tropical monsoon climate; average rainfall: 1,400m-1,600mm; temperature: rises up to 38°C in summer and decrease to 10°C in winter	Older alluvial soils in the adjoining parts and newer alluvial soils along the valley; soils: deep to very deep, sandy clay loam to clay loam texture, brown to grey brown and olive brown colour, high in water holding capacity; more sands near the river beds and more clay further away; air dry moisture between 3.0%-5.8%; slighltly acidic (*p*H: 5.8-6.8); sesquioxide content is higher, N_2 (0.05%-0.1%), P_2O_5 (0.1%-0.6%), K_2O (0.05%-0.1%), organic matter contnt between 0.75%-1.0%	Mainly tropical deciduous; far away locations from the valley, open sals and dense mixed types are common; some indigenous species include Kusum, Mahua, Siris, Arjun etc.	Purely agricultur-at tact, Aman, Aus, Boro, Wheat, Suger-cane, Potatoes are the main cultivated crops; pulses and other vegetables also grown	V E R Y G O O D	II

(*Contd.*)

(Contd.)

Units	Area Delimitation	Geological Foundation	Topographic Conditions	Hydrological Conditions	Climatic Phenomena	Soil Character	Vegetal Cover	Present Land Use	Land Quality/ Potentiality	Land Class
2. Lower Subarna-rekha valley	Dantan I & II Blocks and southern parts of Nayagram (Belda area) Block	Younger alluvium (Dantan-I) as well as older alluvium with sand, silt and clay	Monotonously flat terrain with very low average relief (below 30m) and extremely low relative relief (below 10m); general gradient towards and south and south-east, average slope below 1° (level); low dissection and ruggedness indices; depositional characteristics especially chars (sand bars) in the river channel with broad flood plains	Subarnarekha is the principal river, lateral corrosion, meandering course and broad floodplains; stream frequency: very coarse (below 0.5); drainage density: very low (below 0.5); depth of underground water: <5m, 5m-7m (premonsoon) 1m-3m in the valleys and 3m-5m in the alluvial tract (post - monsoon)	Warm and moist climate (tropical monsoon); monsoon rainfall in summer, average rainfall between 1,500mm-1,700mm, temperature rises upto 37°C in summer and decreases to 10°C in winter, cyclonic disturbances are conspicuous features	Older as well as newer alluvial soils; physical and chemical properties are more or less same like that of the soils of upper Subarnarekha valley	Tropical deciduous species found scatteredly in cultivated fields; important species include Mango, Babla, Siris, Bamboo, Neem etc., fruits and vegetable gardens common nearer the settlements	Fertile agricultural tract, cereal crops (Aman, Aus, Boro, Wheat) are cultivated. Vegetables and pulses are also practised in the flood plains	E X C E L L E N T	I

(Contd.)

(*Contd.*)

Units	*Area Delimitation*	*Geological Foundation*	*Topographic Conditions*	*Hydrological Conditions*	*Climatic Phenomena*	*Soil Character*	*Vegetal Cover*	*Present Land Use*	*Land Quality/ Potentia-lity*	*Land Class*
3. Keleghai Basin	Major area of Narayangath and parts of Keshiary	Mainly older alluvium of Quatern-ary period containing sand, silt, clay and gravels	Gently slopping flat plain with low (30m-60m) average relief, moderately low relative relief, average slope: almost level (below 1°) and in some parts very gentle (1°-2°), river bank erosion common; flooding in the rainy season results in flood aluviation; moderate to moderately high dissection index; low ruggedness index	Keleghai—the only river with narrow channel and sharply cut banks; depth of underground water: 5m-7m and above (pre-monsoon); 3m-5m (post-monsoon); stream frequency: very low (below 0.5) to low (0.5-1.00); drainage density is very low (below 0.5) to low (0.5-1.00)	Modified tropical monsoon climate; temperatures, rainfall, and other climatic conditions more or less same like other units of gently sloping flat plain	Combination of lateritic and older alluvial soils; physical and chemical characteristics more or less same like other units having above mentioned soils; organic matter content is comparatively high than the lateritic soils of the plateau fringe	Mainly tropical deciduous	Purely agricultural tract; local and H.Y.V. rice, vegetables are grown; forest is present in parts of Keshiary	G O O D	III
4. Murli Basin	South-western parts of Nayagram Block	Older alluvium (laterite)	Gently sloping plain with low average relierf (30m-60m) and moderately low relative relief (10m-30m); average slope: level (below 1°) to very gentle (1°-2°); river bank erosion common; moderately high dissection index and low ruggedness index	Murli is a Khal (rivulet); pre-monsoon water table depth between <5m and 5m-7m; stream frequency: very low (below 0.5) to low (0.5-1.00); drainage density: low (below 0.5)	Modified tropical monsoon climate; climatic conditions similar like the former three units	Lateritic soils; soil characteristics are same as units 2 of plateau fringe	Tropical deciduous species that include Sal, Babla, Kusum, Mango etc.	Mainly agricultural tract with considerabl e forest cover	F A I R	IV

Keshiary-Bhasra Area

Bounded by the latitudes of 21°1′ N-21°10′ N and longitudes of 87°10′E-87°15′E the Keshiary-Bhasra area (Keshiary Block) is situated at the northern part of the area under study. The area represents a dual character of undulating terrain and aggradational plain. General geology belongs to both older and newer alluvium. It is characterized by warm and moist climate (tropical monsoon) with cyclonic disturbances. Drought in the undulating terrain and flood in the reverine floodplain are the regular phenomena. The undulating sections dominated by forest cover whereas in the upper and lower flood plains, scattered patches of deciduous species are common. It covers an area of 32.67 sq. km. The Keshiary-Bhasra area has been divided into three geomorphic units (Jana, 1998) (Figs. 11.15, 11.16 and 11.17; Table 11.3).

Sorrong-Anikola Area

Bounded by the latitudes of 21°48′ N-21°54′ N and longitudes of 87°17′ E-87°22′ E, the Sorrong-Anikola area (Dantan-I Block) is located in the south-eastern part of the study area. Covering an area of 32.45 sq. km it is a part of the gently sloping flat plain. The general geology is alluvium (newer). It is characterized by modified tropical monsoon climate with occasional cyclonic disturbances. No such forest cover is present but tropical deciduous in scattered patches are common. Based on the photo pattern the Sorrong-Anikola area has been divided into three geomorphic units as follows (Jana, 1999) (Figs. 11.18, 11.19, and 11.20, Table 11.4).

SECTION III: LAND EVALUATION AND CLASSIFICATION

Land Evaluation

Land evaluation can fruitfully be made by a thorough assessment of the land potentiality. Accordingly, for land evaluation purposes the geomorphological attributes are rated (Table 11.5).

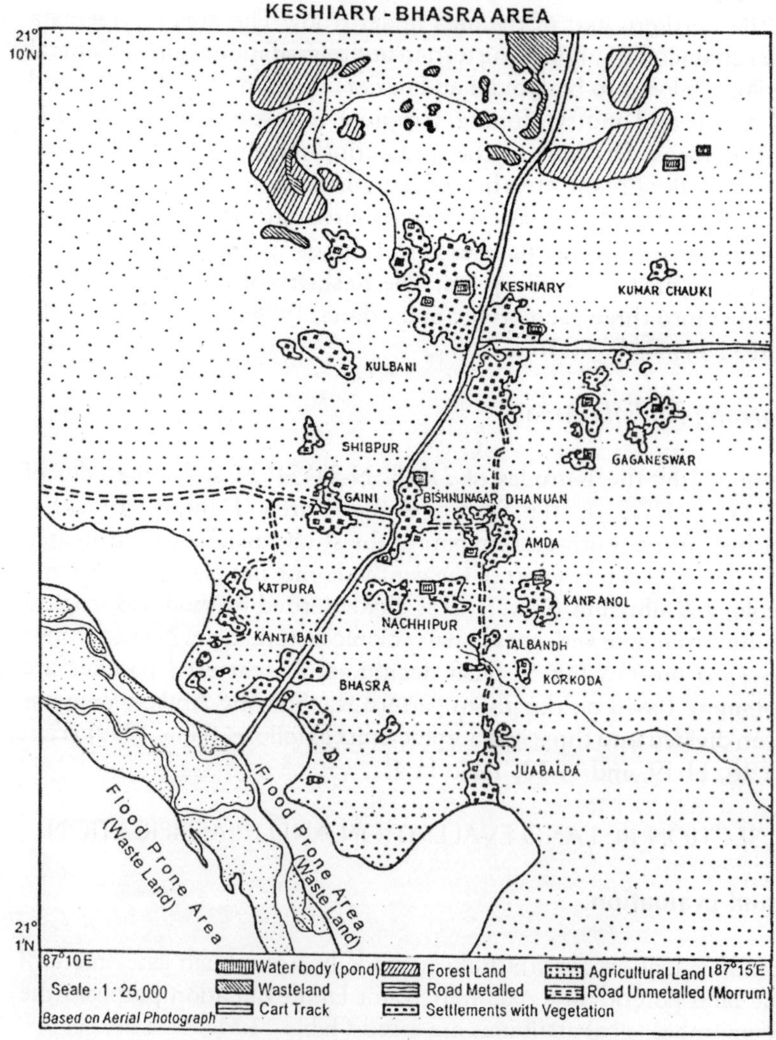

Figure 11.15: Keshiary-Bhasra Area

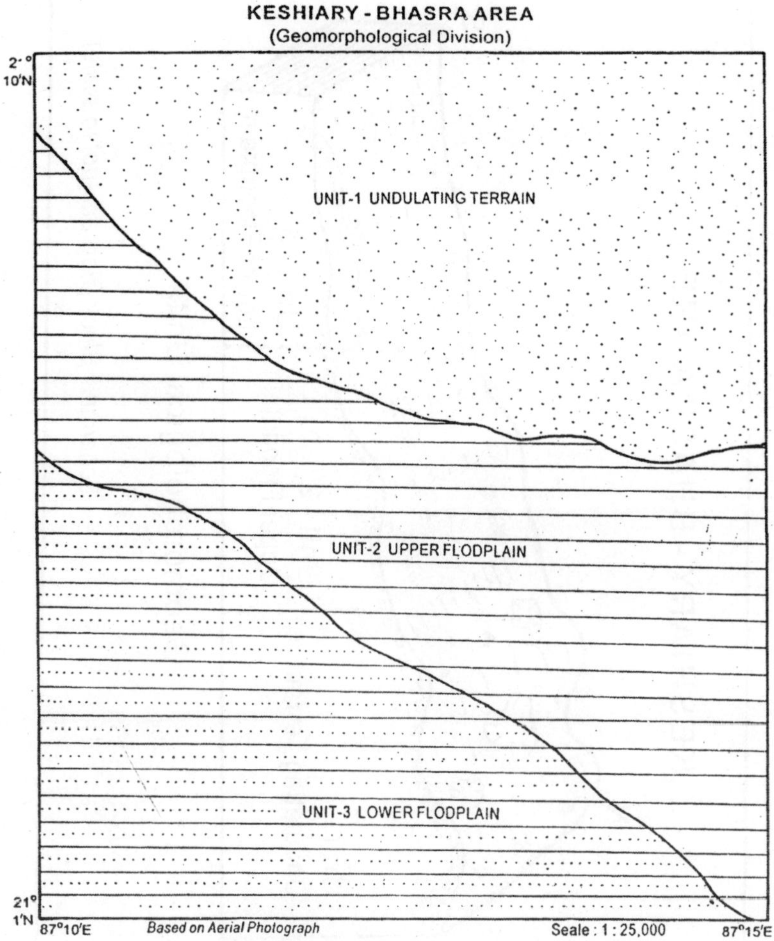

Figure 11.16: Keshiary-Bhasra Area (Geographical Division)

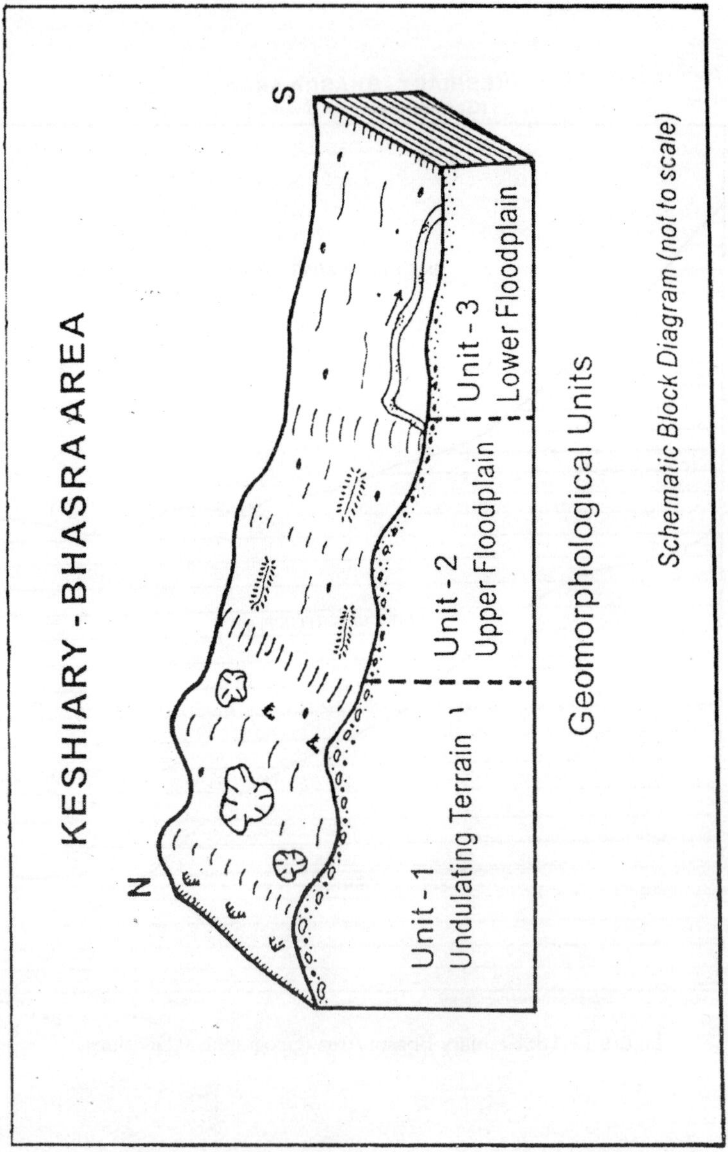

Figure 11.17: Keshiary-Bhasra Area: Geomorphological Units

SORRONG - ANIKOLA AREA

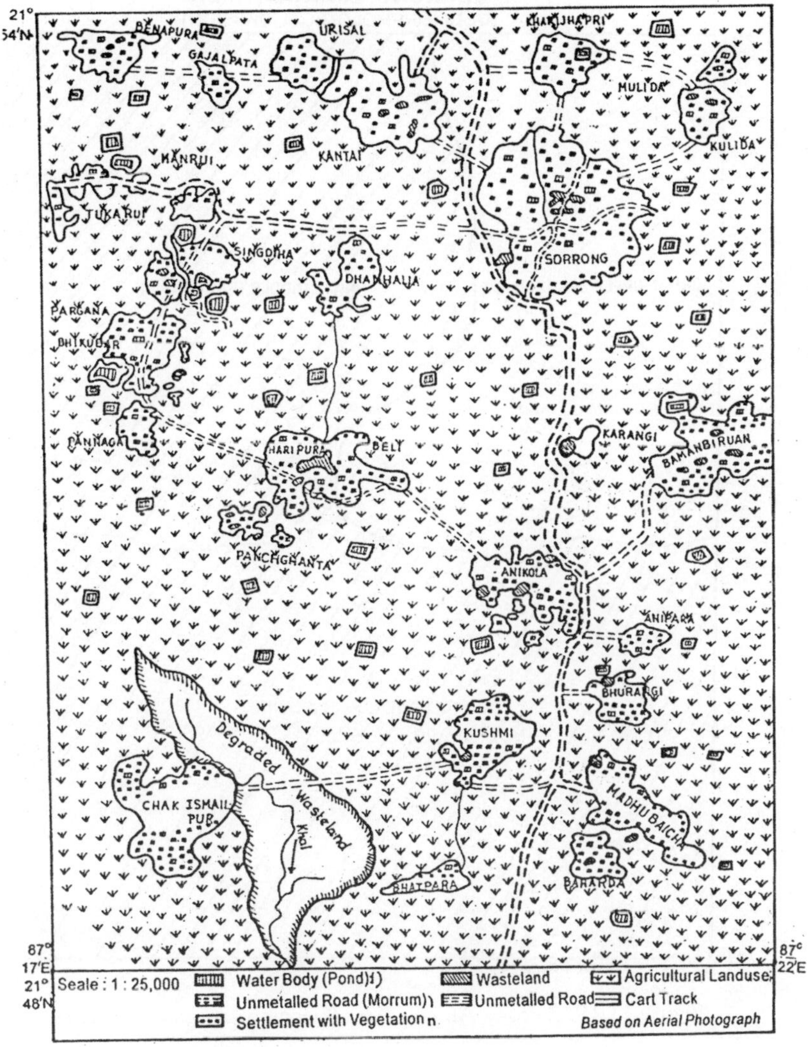

Figure 11.18: Sorrong-Anikola Area

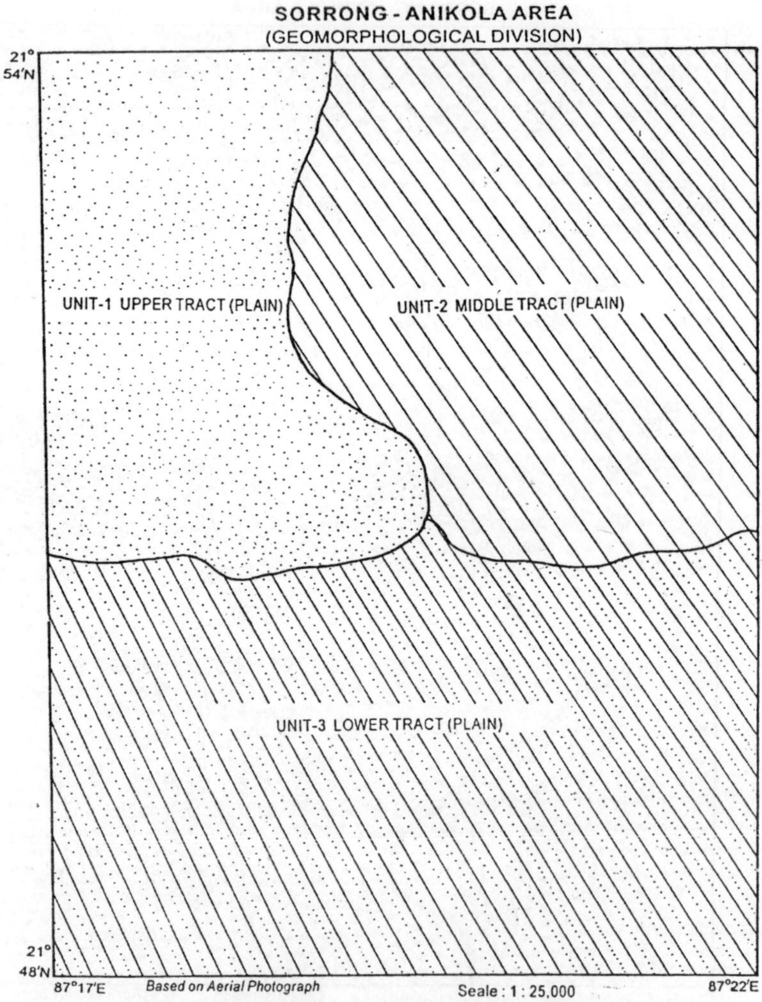

Figure 11.19: Sorrong-Anikola Area (Geomorphological Division)

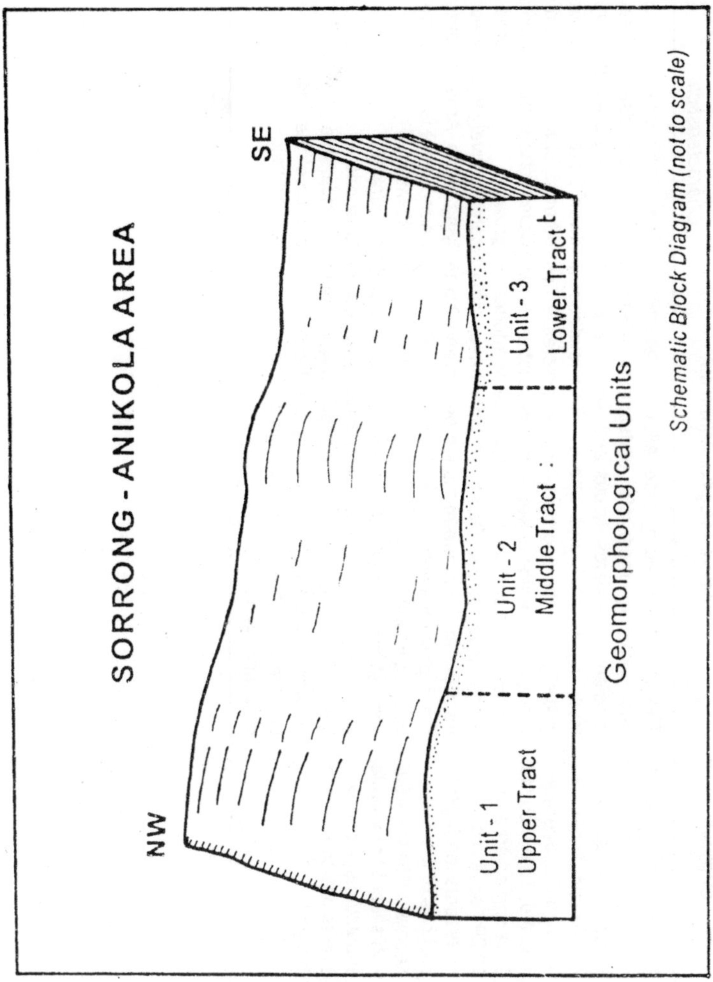

Figure 11.20: Sorrong-Anikola Area

Table 11.3: Keshiary-Bhasara Area

Geomorphic Units	*General*	*Geomorphological Attributes*	*Land Use Characteristics*	*Geomorphology and Land Use*
Unit-1: Undulating Terrain	This land unit includes Keshiary (J.L. No. 106), Kulbani (J.L. No. 87), Kumar chauki (J.L. No. 64), and Gaganeswar (J.L. No. 154) mouzas. Covering an area of 10.291 sq. km it represents the higher sections of the Keshiary-Bhasra area with slight to moderate undulations. Reserve forests are common in this unit.	The unit is covered by older alluvium (lateritic). It has slight to moderate undulations. General gradient is towards south and south-east with level slope (30′ and above). The surface is gravelly, pisolitic and nodular in character. It has moderate average relief and low relative relief, moderate dissection index and low to moderate ruggedness index. Depth of watertable varies from 5m to 7m in pre-monsoon period and 3m to 5m in post-monsoon period. Soils are lateritic (dry and completely bleached) in which iron concretions presents. Soils are coarse textured, poorly aggregated, acidic and have very low to low organic matter content and low water holding capacity. Leaching and gullying are the common phenomena. The unit shows low soil fertility and poor land quality.	The unit shows degraded forest with thick growth of dwarf Sal and other shrubs. Cereal crops are grown in comparatively low-lying areas. Aman (mono crop) is cultivated in favourable locations. A considerable portion of the unit is lying fallow	Because of the unfavourable morphological characteristics of the land/soil the unit is cultivated with mono crop in low lying areas. Degraded forest is found in the undulating terrain.

(Contd.)

(Contd.)

Geomorphic Units	General	Geomorphological Attributes	Land Use Characteristics	Geomorphology and Land Use
Unit-2: Upper Floodplain	This unit includes Shibpur (J.L. No. 118), Gaini (J.L. No. 137), Bishnunagar, Dhanuan (J.L. No. 147), Amda (J.L. No. 159), Kanrarol (J.L. No. 160), Nachipur (J.L. No. 146), Talbandh (J.L. No. 162), and Korkoda (J.L. No. 168) mouzas. Covering an area of 10.58 sq.km. the unit represents an aggradational plain (upper section of the flood plain)	This unit is covered by older alluvium and represents gently undulating topography. General gradient is towards south and slope ranges between 30′ and below 1′. It has low to moderate average relief and extremely low relative relief, low to moderate dissection index and low reggednes index. Depth of water table varies between 4m-6m (pre-monsoon) nd 2m-4m (post-monsoon). Soils are deep to very deep (older alluvial), brown to grey brown, medium in soil texture, slightly acidic, high in organic matter content and water holding capacity. The unit is high in soil fertility and hence a very good quality land.	It is purely an agricultural tract (double cropping mainly and multiple cropping at some places). Aman (local and H.Y.V.), Boro, Wheat, Potatoes are the main cultivated crops. Pulses and other vegetables are also grown significantly.	Due to more favourable geomorphological conditions the unit represents high soil fertility and a very good quality land reflecting high amount of crop production.
Unit-3: Lower Floodplain	This includes Katpura (J.L. No.), Kantabani (J.L. No. 133), Bhasra (J.L. No. 139) and Juabalda (J.L. No. 167) mouzas. Representing an aggradation plain (lower section of the foodplain) the unit covers an area of 11.787 sq. km.	The unit is covered by newer alluvium. It is a monotonously level plain. Renewal of the floodplain by the siltation due to flood is a regular phenomena. Depositional features especially the 'chars' (sand bars) in the channel is common. Slope is towards south and south-west and varies between 20′ and 30′. It has very low average relief and extremely low relative relief, low dissection and ruggedness indices. Depth of water table varies between <5m (Pre-monsoon) and 1m-3m (Post-monsoon). Soils are newer alluvial, medium to fine texture, brown to grey brown in colour, slightly acidic, high in water holding capacity and organic matter content. Soils are much more fertile due to additional siltation and accordingly excellent quality of land.	It is a fertile agricultural tract (double and multiple cropping practised). Aman (local and H.Y.V.) and Boro paddy, wheat, sugarcane, potatoes are cultivated mainly. Pulses and vegetables are grown significant near the River Subarnarekha.	Traversed by the River Subarnarekha the unit represents excellent quality of land/soil and hence bumber Crop production.

Table 11.4: Sorrong-Anikola Area

Geomorphic Units	General	Geomorphological Attributes	Land Use Characteristics	Geomorphology and Land Use
Unit-1: Upper Tract	This unit includes Urisal (J.L. No. 161), Benapura (J.L. No. 164), Gajalpata (J.L. No. 210), Manrui (J.L. No. 167), Tukarui (J.L. No. 168), Singdiha (J.L. No. 208), Paragana (J.L. No. 209), Bhikubarh (J.L. No. 207), Pannaga (J.L. No. 206), Haripura (J.L. No. 216), Beli (J. L. No. 215), and Panchghanta mouzas. It is relatively higher section of the Sorrong-Anikola area and covers in area of 7.13 sq. kms.	It is mainly covered by recent alluvium underlained by sandstones at places. The terrain is flat with level slope (30′) slightly higher than the other two units. It has low average relief and extremely low relative relief, low dissection and ruggedness indices. Soils are deep, brown to grey brown, light to medium in texture (sandy loam to clay loam) and are acidic (slightly to moderately) with high organic matter content. Ground water level ranges between <4m (pre-monsoon period) and 1m-2.5m (post-monsoon period). The land/soil has high water holding capacity and soil fertility-good quality land.	It is purely an agricultural tract-monocropping and double cropping in places. Aman (both local and high yielding). Aus and Boro are the cultivated cereal crops. Vegetables are grown locally.	Because of the favourable morphological characteristics of land/soil the unit is cultivated with cereal crop (monocrop). The crop production is quite satisfactory.
Unit-2: Middle Tract	This unit includes Kantai (J.L. No. 160), Kharijhapri (J.L. No. 158), Kulida (J.L. No. 223), Mulida (J.L. No. 224), Sorrong	Covers by newer alluvium it is a monotonously flat terrain with general gradient towards south and south-east (level slope—25′). It has low average relief, low dissection and ruggedness indices. The soils are deep to very deep (alluvial) and have clay to clay loam and sandy clay loam (medium) in texture.	The unit is a very good agricultural tract (double cropping practised). Aman (local and H.Y.V.) and Boro are the main cultivated cereal crops.	The morphological characteristics of land/soil being more favourable the unit is very fertile and reflects

(Contd.)

(Contd)

Geomorphic Units	*General*	*Geomorphological Attributes*	*Land Use Characteristics*	*Geomorphology and Land Use*
				high amount of production.
Unit-3: Lower Tract	Dhanhalia (J.L. No. 212), Bamanbiruan (J.L. No. 218), Karangi (J.L. No. 214) mouzas. It covers an area of 10.10 sq. km and represents comparatively lower elevation than unit-1.	It is brown, grey brown and olive brown in colour. The groundwater level here varies between 3m (premonsoon period) and 1m-2.25m (post-monsoon period). In general soils are slightly acidic as well as high in moisture and organic matter content. Soil fertility being high, the unit represents a very good quality land.	Vegetable growth is not so significant here but Kitchen garden is however present in almost all the households.	
	This Land unit includes Anikola (J.L. No. 217), Anipara (J.L. No. 335), Kushmi (J.L. No. 201), Bhurangi (J.L. No. 336), Madhubaicha (J.L. No. 337), Baharda (J.L. No. 338), Bhatpara (J.L. No. 197), and Chak Ismailpur (J.L. No. 199) mouzas. covering an area of 15.22 sq. km. the unit shows the characteristics of level plain with isolated depressions.	The unit is covered with newer alluvium. It is monotonously flat plain with gradient towards south and south-east (level slope of 20'). It has low average relief, extremely low relative relief, low dissection and reggudeness indices. The groundwater condition is better here (2.75m in pre-monsoon and 1m-2m in post-monsoon periods). The soils are very deep, fine textured, grey brown to olive brown in colour and slightly acidic. Soil erosion (especially canal bank) is common. Flood during rainy months due to excessive discharge of water in the canal is a regular phenomena. Fresh yearly deposits of silts are common in the area adjacent to canal. The unit has very high water holding capacity, soil moisture content, organic matter content and soil fertility and accordingly it has excellent quality of land/soil.	It is purely an agricultural tract (mainly double cropping and multiple cropping at places). Aman (local and H.Y.V.), Boro, Wheat (at some places) are the major cultivated crops. Vegetables are grown in the area adjacent to canal Kitchen gardening is an essential part in all the households	This unit has the excellent quality of soil and land characteristics. Being traversed by the canal it has the facility of having irrigation and accordingly has bumper crop production.

Table 11.5: Geomorphological Attributes and their Rating Values

Sl. No.	Attributes	Description	Rating Values
1.	Average relief	Very low	100
		Low	80
		Moderate	60
2.	Relative relief	Extremely low	100
		Moderately low	80
		Low	70
3.	Average slope	Level	100
		Very gentle	90
		Gentle	80
4.	Depth of water table (post-monsoon)	1 m-3m	100
		3m-5m	60
		> 5m	40
5.	Dissection index	Low	100
		Moderate	70
		Moderately high	60
6.	Ruggedness index	Low	100
		Moderate	60
7.	Stream frequency	Very low	100
		Low	80
		Medium	60
		High	40
8.	Drainage density	Very low	100
		Low	80
		Medium	60
9.	Erosion	No erosion	100
		Soil erosion (normal)	90
		River-bank erosion	40
		Leaching	40
		Gullying	40
10.	Soil:		
	• Texture	Medium	60
		Coarse	40
		Very coarse	30
	• *p*H	Slightly acidic	90
		Slight to moderately acidic	70
		Moderately acidic	65
	• Organic matter	High	100
		Medium	80
		Low	60
		Very low	30

Land Classification

Land classification is a scientific appraisal of the physical characters of the land, inherent soil qualities etc. The main objective of land classification is to understand potentiality, capability and suitability for its optimum utilization. Broadly speaking, the general purpose of land classification is to provide a systematic basis for the study of crop and soil relationship so as to increase the productivity of land by various means. There is an urgent need for such studies because per capita supporting capacity of land is decreasing with increasing population. Land capability measurement offers a scientific judgement for the conservation of land under specific ecological conditions.

Land classification in one hand helps to find out efficiency of land for a particular use and on the other it helps to prevent its improper use leading to erosion hazards and deterioration of land quality. So, for the sake of optimum productivity, the capability of each and every bit of land should be measured considering its inherent pedo-geomorphic characters, as well as limitations due to environmental hazards. Moreover, capability classification enables the farmers to use the land properly for sustainable production under required management measures.

In the present context different land units of the area under study are categorized/graded as class I, II, III, IV and V on the basis of their potentiality, capability and degree of limitation to land use (Table 11.6).

- *Class I Land (Excellent)*: This type has excellent qualities and has favourable slope, relief, drainage, soil texture, soil reaction and high organic matter content. Such land requires ordinary management practices. It is suitable for intensive cultivation as well as for wide range of crops. The lower Subarnarekha valley falls under this category.
- *Class II Land (Very good)*: This category of land has very good qualities and has favourable aspects of relief, slope, drainage, soil texture and organic matter content with slight limitation in regard to riverbank erosion. It is also suitable for intensive cultivation with ordinary to fair management practices. The upper Subarnarekha valley falls under this category.
- *Class III Land (Good)*: Land with favourable aspects of relief, slope, drainage, soil texture and chemical capacity. It reflects level to very gentle slope, favourable water conditions and fairly high productivity. This type of land is suitable for intensive cultivation. It needs fair management practices and is suitable for most crops. The Keleghai basin falls under this category.

Table 11.6: Land Evaluation and Land Classification Based on Rating Values

Geomorphological regions	*Units*	*Average relief*	*Rating value*	*Relative relief*	*Rating value*	*Average slope*	*Rating value*	*Depth of water table*	*Rating value*
(1)	(2)	(3)	(4)	(5)	(6)	(7)	(8)	(9)	(10)
Plateau fringe	1. Upper degradational surface	Moderate	60	Moderately low	80	Gentle	80	3m-5m	60
	2. Lower degradational surface	Do	60	Low	70	Very gentle to gentle	85	3m-5m	60
Gently sloping flat plain	1. Upper Subarnarekha valley	Low	80	Extremely low	100	Level to very gentle	95	1m-3m	100
	2. Lower Subranarekha valley	Very low	100	Do	100	Level	100	1m-3m	100
	3. Keleghai basin	Very low to low	90	Moderately low	80	Level to very gentle	95	3m-5m	60
	4. Murli basin	Low	80	Moderately low	80	Level to very gentle	95	3m-5m	60

Dissection index	*Rating value*	*Ruggedness index*	*Rating value*	*Stream frequency*	*Rating value*	*Drainage density*	*Rating value*	*Erosion*	*Rating value*	*Soil texture*
(11)	(12)	(13)	(14)	(15)	(16)	(17)	(18)	(19)	(20)	(21)
Moderate	70	Moderate	60	Low	80	Very low to low	90	Gullying	40	Very Coarse
Moderate to moderately high	65	Low to moderate	80	Medium to high	50	Low to medium	70	Leaching, gullying	80	Coarse
Low	100	Low	100	Very low to low	90	Very low to low	90	River bank erosion	40	Medium
Low	100	Low	100	Very low	100	Very low	100	No erosion	100	Medium
Moderate to moderately high	65	Low	100	Very low	100	Very low to low	90	River bank erosion	40	Coarse
Moderately high	60	Low	100	Very low to low	90	Low	80	River bank erosion	40	Coarse

(Contd.)

(Contd.)

Rating value	Soil pH	Rating value	Organic matter	Rating value	Sum of Rating value	Land quality/potentiality	Land class
(22)	(23)	(24)	(25)	(26)	(27)	(28)	(29)
30	Slightly to moderately acidic	70	Very low	30	750	Poor	V
40	Moderately acidic	65	Low	60	785	Poor	V
60	Slightly acidic	90	High	100	1045	Very good	II
60	Slightly acidic	90	High	100	1150	Excellent	I
40	Slightly to moderately acidic	70	High	100	930	Good	III
40	Slight to moderately acidic	70	Low	60	855	Fair	IV

Note : Rating Values Based on R.E. Storie (1933).

- *Class IV (Fair)*: The Murli basin is included in this category. Here land category is moderate with some limitations in regard to relief, slope, drainage, soil texture and chemical capacity. It is a medium quality land with less wide range of suitability than class I, II and III and require good management practices. This land type may give good results with certain specialized crops.
- *Class V (Poor)*: The upper and lower degradational surfaces of the plateau fringe represent poor quality of land. It has limitations in terms of relief, slope, drainage, soil texture and chemical capacity etc. The land here is characterized by gently undualating slope, unfavourable water conditions, occasional natural hazards etc. and has medium to poor productivity. According it has limitations in agricultural practices.

SECTION IV: PLANNING STRATEGIES FOR MACRO AND MICRO REGIONS

1. Macro Regions

Reflecting dichotomy of undulating lateritic tract and flat alluvial terrain the study area exhibits gully erosion, flood hazards, drought condition, fluctuating underground water etc.

In the undulating uplands, degraded land surfaces are gradually increasing in dimension through gully formation due to massive deforestation, improper water management and in some cases irrational use of land. The intensity of soil erosion mainly through rills and gullies spread over large area and gradually it encroaches into agricultural lands.

The undulating topography with light soil texture; a number of rivers, rivulets and canals passing through the area also accelerate soil erosion. The western plateau fringe is more prone to soil erosion hazards as compared to the monotonously flat plain of the east. The area is generally not affected by floods, but unusual torrential spell of rainfall in certain years result in high flood leading to severe damage to crops and properties (Jana, 1999).

The meteorological data of Midnapore and Jhargram reflect high probability of droughts that results in agricultural drought-like conditions (Das and Sen, 1990). Thus, the occurrence of dry spell and quantity of rainfall during the growing period of paddy dictate the success or failure of agriculture and gives important guidelines to planners and decision makers to optimize the use of water through timely irrigation schedules (Sivasamy and Bhattacharya, 1979).

Keeping in view with such nature of the area and its existing land use practices, the planning strategies for prospective land uses are suggested (Table 11.7).

2. Micro Regions

Micro-level land use planning and management for sustainable development is a need of the hour. The primary aim of micro-level land use planning is to improve the socio-economic standard of rural downtrodden, to maximize land productivity, to use properly the unemployed rural manpower and provide other basic needs for the local communities (Jana, 1998).

As agriculture is the main pillar of economy, it requires more careful investigation of local pedo-geomorphic characteristics and problems as well as integration of various land use patterns related with agricultural activity for the effective planning of rural land use and sustainable economic development (Jana, 1999). The strategies for micro-level land use planning (e.g. Keshiary-Bhasra Area and Sorrong-Anikola Area) are given in Table 11.8.

SECTION V: FUTURE RECOMMENDATIONS

Keeping in view with the contrasting surface configuration and diversified problems of the area concerned the author would like to recommend the following strategies for land use planning in general and sustainable rural development in particular.

- *Optimum Use of Water*: Renovation of existing irrigation system is essential as a critical input to achieve the targeted growth rate of agricultural production. The adoption of measures towards the optimum use of water for augmenting agricultural production are as follows:

  - Planning for consecutive use of surface and ground water resources,
  - Improving water use efficiency through renovation and modernization of existing system,

Table 11.7: Strategies for Macro Regional Land Use Planning

Macro Geomorphic Regions	*Planning Units*	*Major Problems*	*Land Capability*	*Land Class*	*Planning Measures for Prospective Land Uses*
Plateau Fringe	1. Upper Degradational Surface	Riverbank erosion, boulder excavation, intricate gullies, reckless deforestation, fluctuating underground water, low fertility	Poor-limitations in agricultural practices and needs proper management practices	V	Land should not be used for agriculture, only, upper parts can be utilized for dry farming and commercial forestry, agriculture in favourale locations and some parts may be utilized for selected forest-based industries, aforestation in deforested areas.
	2. Lower Degradational Surface	Predominant leaching with gully erosion, soil erosion during rainy months, boulder extraction, massive deforestation, water crisis, low fertility	Poor-limitations in agricultural practices and requires proper management practices	V	Extensive agriculture with proper irrigation facilities, upper parts of the undulation can be utilized for forestry purposes on commercial basis, management of forest resources and development of forest-based industries.
Gently Sloping Flat Plain	1. Upper Subarnarekha Valley	Sharply cut banks with rocky beds, river bank erosion in the upper sections	Very good—suitable for intensive cultivation with ordinary to fair management practices	II	Double and multiple cropping with more irrigation facilities especially river lift, in the upper section agro-forestry can be encouraged.
	2. Lower Subarnarekha Valley	Lateral corrosion, meandering course and broad flood plain, cyclonic disturbances	Excellent—suitable for intensive cultivation as well as for wide range of crops and requires ordinary management practices.	I	Multiple cropping with the adoption of modern agricultural technology, vegetables and pulses can be practised in the flood plains on commercial basis.
	3. Keleghai Basin	Phenomena of river bank erosion, flooding in the rainy season	Good—suitable for most crops and needs fair management practices	III	Double and multiple cropping in suitable locations with technological inputs, forest can be encouraged in the tract having lateritic soils.
	4. Murli Basin	Presence of river bank erosion, lateritic soil, deforestation	Fair—medium quality land with less wide range of suitability and requires good management practices	IV	Emphasis is to be given on agriculture, forestry and forest based industry.

Table 11.8: Strategies for Micro Level Land Use Planning

Micro Geomorphic Regions	*Planning Units*	*Major Problems*	*Land Quality*	*Planning Measures for Prospective Land Uses*
Keshiary-Bhasra Area	1. Undulating Terrain	Dry and completely bleached lateritic soils, low organic matter content, low water holding capacity, common phenomena of leaching and gullying, low soil fertility	Poor	Regulation for illegal felling of trees to check soil erosion, double cropping in favourable locations with the adoption of suitable technology, comparatively higher elevations as well as waster lands may be utilized for forestry and fodder crops rather than agriculture.
	2. Upper Floodplain	No such problems have been observed	Very Good	Double and multiple cropping can be extended on large scale with the adoption of modern technology, vegetables can be grown on commercial basis in the Palanla Khal area, agro-forstry may be encouraged in some places.
	3. Lower Floodplain	No significant problems have been envisaged	Excellent	Multiple cropping can be extended at a large scale with advanced agricultural implements, vegetables and pulses can be grown on commercial basis, social forestry may be suggested on the river bank.
Sorrong-Anikola Area	1. Upper Tract	Slightly to moderately acidic soil, no other significant problems	Good	Application of lime to remove soil acidity, double cropping at a large scale may be practised with the help of irrigation, social forestry may be suggested in wastelands.
	2. Middle Tract	Slightly acidic soil, no other significant problems	Good	Application of lime; multiple cropping with proper irrigation, other technological inputs and management practices; social forestry in localities unsuitable for cultivation.
	3. Lower Tract	Soil erosion in the canal bank, flood due to excessive discharge in canal but no such damage	Excellent	Multiple cropping can be extended at large scale with required irrigation, vegetables and pulses can be grown commercially in the areas adjacent to canal, social forestry on the canal bank to prevent soil erosion.

- Raising water rates and rationalize water use to foster the motivation for economy in water use,
- Involvement of farmers in the management of irrigation,
- Promotion of adaptive research and development to ensure more cost effective as well as efficient execution and management of irrigation,
- Taking steps for greater financial investment from the institutes in the implementation of minor irrigation schemes,
- Completion of ongoing irrigation schemes at war footing,
- Improving utilization of irrigation potential by integrating all functions of irrigated agriculture,
- Encouraging participation of private sector in irrigation system.

Despite the above-mentioned measures for optimum use of water, the complex interconnectedness of water systems demand a holistic (taking catchments and command area development approach) management, which would be based on a balanced consideration (technical, environmental and economic considerations) of the needs of people and environment.

➢ *Land and Soil Conservation*: The measures needed are:

- Anchoring soil to the earth through tree plantation, agro-forestry, farm forestry and enhanced green cover
- Adopting integrated watershed approach for rainfed farming which promotes need based soil conservation measure in arable and non-arable lands.
- Nourishing land through conservation of rain water, recharge of ground water and judicious use of water.
- Incorporating organic wastes and manures into the soil and cropping pattern for the conservation of soil and moisture.
- Storing of water in the suitable places, viz. ponds, tanks or reservoirs that control flood and reduce soil erosion. In addition, the collected water can be utilized at critical periods of cropping.

- *Enrichment of Soil Fertility*: In the study area the soil fertility can be enriched:

    - Introducing vermiculture.
    - Application of chemical fertilizers in measured doses,
    - Application of organic manures and bio-fertilizers in larger amounts.

- *Restructuring Cropping Pattern*: To ensure optimum use of soil moisture, emphasis should be given on the selection of crops having deeper root system and capacity to cope with drought. Moreover, attention is to be placed on the system of intercropping which can provide fodder for animals and pulses for consumption without deteriorating the fertility status of soils.
- *Joint Forest Management*: Joint Forest Management (J.F.M.) is the sharing of products, responsibilities, control and decision-making over forest land between forest department and the local people. In the study area particularly in the lateritic tract J.F.M. is the best prescriptive measure to save dwindling forests, to prevent soil and gully erosion and properly utilize the localized forest resources for the economic development of the region in general and tribal communities in particular.
- *Development of Livestock*: The development of agriculture in the undulating and infertile lateritic upland is characterized by low yield of principal crops. Such situation necessitates the development of livestock through vigorous grassland and pasture development programmes. Livestock farming in combination with agriculture would be an alternative source of economy to the rural households concerned.
- *Promotion of Rural Road*: Rapid development of rural roads is essential for linking village to nearby markets and services. This linkage brings different areas in close contact and interaction between villages and towns is facilitated which has a greater positive impact on rural development in general and land utilization in particular. Thus, an efficient rural road network can accelerate the socio-economic progress of the countryside by:

    - Facilitating rapid movement of rural traffic at reduced cost and mobility of rural people

- Carrying surplus food grains, vegetables and animal products to urban centre.
- Facilitating the introduction of modern inputs, viz. HYV seeds, chemical fertilizers, pesticides, insecticides etc. for bumper crop production.
- Rendering public administration more efficient for effective land utilization.

Concerning rural linkage some sort of progress has been achieved in the study area. But the connectivity is to be promoted for effective land utilization thereby providing ultimate benefits to the rural folk.

- *Rural Electrification*: Rural electrification is one of the most basic needs for the transformation of rural life. It is the key factor in the development of agro-based industries. The agricultural sector consumes major share of electric power as it is used for pump set energisation towards irrigation. It may be pointed out that the W.B. State Electricity Board has extended electricity to some parts in the study area, but frequent load-shedding is a regular phenomena which interrupts agricultural operations. Efficient rural electrification in the remaining part is recommended for effective land use planning and development.
- *Development of Planned Rural Markets*: The development of planned market centres at a certain intervals gives momentum to the rural economy, because it would be an ideal foci for buying and selling of rural agricultural commodities. This type of market centres can be developed at the meeting point of rural roads. The government should encourage and extend their help towards the development of such market centres.
- *Development of Rural Bank*: The credit facilities are essential for optimum land utilization. In the study area a number of rural banks can be set up for meeting the credit requirements of the farmers towards prosperous agriculture.
- *Development of Forest Based Industries*: The situation of drought-like conditions in the undulating lateritic tract necessitates the shifting of emphasis from the agricultural sector to the development of forest-based industries. Owing to the availability of forest resources, viz. Sal leaf, Cashewnut,

Neem fruit, Mahua flower, Babui grass, Kend leaf etc. forest based industries can be set up in large scale with full vigour. The proposed industries include:

- Sal leaf utensils have greater demand in the adjoining region, district towns and Calcutta Metropolis,
- Bidi making by Kend leaf to meet the local needs and demands of the backward region especially the western fringe of West Bengal,
- Neem soap making through the oil extracted from Neem fruit,
- Mushroom cultivation as a substitute of protein diet as well as for medicinal purposes,
- Scientific collection and processing of honey on commercial basis,
- Making of sports accessories using Bamboo and other woods,
- Establishment of more paper mills by the use of available Bamboo.

It may, however, be pointed out that the development of forest based industries needs commercial exploitation of forest resources. Such attempt without management practices results in irrational deforestation and environmental degradation. So, the rational management of forest resources is essential for restoring ecological balance and maintaining constant flow of raw materials to the forest-based industries.

➢ *Development of Eco-Tourism*: Tourism, among other human activities, today has been recognized as a fastly expanding potential industry. As a smokeless industry with insignificant environmental impact it can be developed with an aim to satisfy man's aesthetic needs to safeguard human health and to create more employment opportunities for the youths. Representing dual character in relief the study area has a greater scope for developing eco-tourism as an alternative economic paradigm. The forest enriched lateritic tract with high quality drinking water attracts many tourists on health grounds, which will encourage the development of health resorts.

➢ *Training Programme*: For the use of modern agricultural

implements in land utilization, proper training of farmers is an utmost necessity. Training of agricultural labourers is also needed to increase their field efficiency. Agricultural training programmes can give a fresh lease to the local farmers in the development of agro-based economy. The training programmes include training camp, short demonstration camp, field visit, *kisan mela*, training in maintenance and repairing of agricultural instruments etc. which more or less supports introduction of modern methods in agricultural practices.

The radio, television and video systems are very effective media by which a large number of farmers can be approached with up-to-date technical know-how quickly and efficiently.

SECTION VI: CONCLUDING REMARKS

In the vastly populated country like India, pressure on land is a matter of great concern. For the lack of ecological perception among the people, over-use, mis-use and under-use of land are the common phenomena. That is why proper understanding, actual measurement of land potentiality and required management practices are a major field of research to the geographers.

Exhibiting dichotomy of agriculture and forestry, the study area needs proper development initiatives at government level to optimize land utilization to form a sound sustainable agro-forestry based economy.

In the area under investigation several measures have been suggested to eradicate the identified problems towards effective land utilization. The government has also taken some initiatives in particular. It may be pointed out in this context that an important barrage project has already been designed on the River Subarnarekha expecting to provide a fresh lease to agriculture. The alluvial tract of the east (Narayangarh, Dantan I & II Blocks) will be benefited from the said project, whereas the undulating lateritic terrain (Nayagram, Gopiballavpur-I & II Sankrail Blocks) suffering from water crisis and prolonged drought have not been incorporated in the proposed irrigated area of the project. The author does not find any reason (whether technical or environmental) for the non-inclusion of these blocks and accordingly suggests that the Government. can refresh and rechalk the scheme so that the drought prone areas may be benefited.

There are, however, some minor irrigation schemes (few completed, few ongoing and few proposed) in the western lateritic tract. These are quite insufficient for optimum land utilization which requires to be integrated with the main Subarnarekha Project for long term benefit.

Generally, irrigation schemes (mainly canal) are financed and maintained by the government on subsidy basis imposing a very little tax on the farmers. Long experience shows that the farmers are not ready to pay the taxes and in most cases the yearly tax collection remains beyond the target. Due to increasing price hike and financial stringency the government started to close the provision of giving subsidy to the irrigation schemes. Irrigation experts feel and the author suggests that participatory irrigation management is the only solution wherein the responsibility of maintaining irrigation sources will be imposed on the local farmers. This type of irrigation management has been introduced in Andhra Pradesh, Maharashtra etc., but the Government of West Bengal has made no such attempt till today. It may, however, be pointed out that more and more cultivable lands are being included under water seeking H.Y.V. If new irrigation schemes are not introduced and the existing ones are not renovated by the government there will be chances of crop failure in future years.

Although some measures have been adopted by the Drought Prone Area Programme section, Jhargram towards betterment of agriculture, but more emphasis is to be given in this regard. Keeping in view with the luxuriant forest resources of the undulating upland planned efforts should be made for economic benefit and ecological restoration.

Despite some sort of socio-economic progress, the study area is still backward. It is not beyond doubt that the government initiatives already taken are insufficient to give effective result to land utilization.

During the course of entire study the necessary data and relevant information, which the author his consulted, are duly supplemented by thorough fieldwork. It may especially be pointed out that the study area is the native place of the author that provides an added advantage to him to carry out the work in a proper perspective.

Land is an ultimate asset particularly in a developing country like India where economy is rooted in agriculture. India lives in the villages, so proper land utilization can only strengthen the rural economy which needs an effective integration of physical and human eco-systems approach to land use planning. Strictly speaking, the results of land capability assessment should essentially be integrated with the farmer's preferences and capabilities (Sundram, 2003).

In the present paper, the author has assessed the impact of geomorphic attributes on existing land use and accordingly suggested planning strategies mainly at the macro-regional levels from the geographer's viewpoint. P.R.A. (Participatory Rural Appraisal) method has not been taken into account in the formulation of land use planning strategies. The author has limited his work to the evaluation of geomorphic attributes. From this point there is scope to proceed further towards effective land use planning at micro level.

Despite the above limitation, the present work is considered not only as a scientific one, but also precise, accurate, easy, less time-consuming and at the same time this can safely and satisfactorily be applied to any scale of working for the regional development whether larger or small, under-developed or developed.

REFERENCES

Brunsden, D. (1985). "Geomorphology in the Service of Society", *The Future of Geography*, ed. by R. J. Johnston, Methuen, London & New York, p. 225.

Clark, A. N. (1985). *Longman Dictionary of Geography: Human and Physical*, Longman Group Limited, UK, pp. 29, 337.

Das, C. and Sen, P. K. (1990). "Impact of Deforestation and Geomorphic Forms and Process in Jhargram Forest Range", *77th Indian SC Congress*, Cochin.

Dayal, P. (1990). *A Textbook of Geomorphology*, Shukla Book Depot, Patna, pp. 742-744.

De, N. K. and Jana, N. C. (1997). *The Land: Multifaceted Appraisal and Management*, Sribhumi Publishing Co., Kolkata. 16-31, 45-46, 146.

Fairbridge, R. W. (1968). *The Encyclopedia of Geomorphology*, Ed. Vol. Dowden, Hutchinson & Ross Inc., Stroundsburg, Pennsylvania, p. 420.

Jana, N. C. (1998). "Use of Aerial Photographs for Micro Level Land Use Planning: An Approach to Sustainable Development", *Indian Cartographer*, Vol. 18, December 1998, Kolkata, pp. 101-104.

Jana, N. C. (1998). *Geomorphology and Land Use in South-Western Midnapur, West Bengal*, unpublished Ph.D. Thesis, Burdwan University.

Jana, N. C. (1999). "Application of Remote Sensing Techniques for Rural Land Use Planning and Development", Seminar Proceedings on *Information Technology for Rural Development* (IT Rural 1999), Seminar Organised by USIC, Burdwan University.

Jana, N. C. (1999). "Prospect of Forest Based Industries in Jhargram Sub-

Division of Medinipur District, West Bengal through Sustainable Management of Forest Resources", *Indian Journal of Geography & Environment*, Vidyasagar University, West Bengal.

Sivasamy, K. S. and Bhattacharya, N. (1979). "Droughts in Dry and Moist Sub-Humid Climate Regions—A Case Study of Bankura and Midnapore Districts in West Bengal", *Drought Prone Areas of India*, ed. by N. B. K. Reddy, Published by Rayalaseema Geographical Society, Sri Venkateswara University, Tirupati, pp. 233-236.

Storie, R. E. (1933). "An Index for Rating the Agricultural Value of Soils", Berkeley, California University Agri. Experi. Station, *Bulletin* 556.

Sundaram, K. V. (2003). "Integrating Land Capability with Farmer's Capability: Land Use Planning Approach for the 21st Century", *Occasional Papers*, Bhoovigyan Vikas Foundation, New Delhi.

Verstappen, H. Th. (1983). *Applied Geomorphology: Geomorphological Surveys for Environment Development*, Amsterdam Elsevier, p. 226.

Vink, A. P. A. (1975). *Land Use in Advancing Agriculture*, Advance Series in Agricultural Sciences I, Springer Verlag, Berlin-Heidelberg.

III
Sustainable Livelihoods and Agriculture

12

The Jana Utthan Approach: BAIF's Approach to Poverty Alleviation in Rural Areas

ANIL P. TAMBAY and I. P. BHAGWAT

Challenges in Rural Poverty Alleviation

Despite efforts made over the past few decades, rural poverty in India continues to be significant. While the government and other agencies have strengthened their anti-poverty programmes, and in percentage terms poverty levels have reduced (from 56.4% in 1973-74 to 37.27% in 1993-94), the numbers of the rural poor have remained static. There is realization, however, that the need is to create opportunities for self-employment, for they alone can provide income to the rural families on a sustainable basis.[1]

But poverty is not only an economic phenomenon. It is also deeply embedded in the *socio-political* scenario of an area, it is about *access* of the poor to resources and services, it is about *health, gender equity* and it is also about *empowerment*. There is also the issue of *time* required by families to come up above the poverty line, as is the issue of the *costs* involved in the process. All these learnings have been gained and encapsulated through national/international experiences.

Thus, apart from creating self-employment opportunities, there is a need for poverty alleviation programmes to simultaneously address a host of complex issues. It may be argued that various sectoral programmes do address all these issues in a given area, and the need is for a sincere practice of intersectoral coordination. It may also be argued that Panchayati Raj Institutions (PRIs) are being strengthened in rural areas, to empower the people.

The points are well accepted, but remain as yet a theoretical issue for most Poverty Alleviation Programmes, whether governmental or otherwise. The need, on the other hand, is for a single programme to address the issues leading to poverty alleviation.

Such an approach is being evolved by BAIF, as part of a BAIF-European Union (EU) Project, in 11 Districts across 5 States. This approach, called "Jana Utthan" (People's Upliftment), evolved as part of the Project, and will be briefly described in the following sections.

What is the "Jana Utthan" Approach?

It is a "learning approach" to development, which is flexible, can be used to target the poorest and yet goes beyond being only a "livelihoods approach". This approach has been developed through an on-going European Union sponsored project that is being implemented by BAIF.

The origin of this approach lies in the realization based upon practical experience that the needs of the poorest rural families are diverse and cannot easily be addressed through a standardized model or package of interventions to be applied to all families. It was also realized that the 'poor' are not necessarily a homogenous economic or social category, and differentiating amongst them is a necessity for the better addressing the situation on the ground, hence the W1-W4 categories followed in this project.

While this approach is still characterized as being 'in evolution', the following are some of its *main steps*:

- Rapport building and "laying a dynamic foundation" through area based interventions like community health, water resource development, livestock development, entry point activities etc.
- Village community defines poverty in their own context and identifies the poorest families.

- Village community representatives (from People's Organisations) and each of the poor families together prepare a family Micro Level Plan of economic interventions on the basis of the family's present resources.
- Micro Level Plan aims at assuring food security and raise the family's economic status above the poverty line sustainably.
- Micro Level Plans continuously evolve on the basis of each family's needs, successes and failures. They often include more than one economic intervention.
- People's Organisations constituted by these families, support the family Micro Level Plans by managing input supplies, training, services and monitoring the progress of each family.
- People's Organisations directly link up with mainstream organisations, e.g. government departments, banks, markets etc. to sustain the development process for the community as a whole.

Some salient features of the Jana Utthan Approach include: Reaching the very poorest; Flexible, need-based problem solving approach; Package of interventions with each family to focus on food security, livelihoods and quality of life; Encouragement of individual empowerment and growth, while building on family and community assets; Family specific approach evolves into community development approach; Community concerns—gender, health, education etc. addressed; and Community ownership of the development process becomes a practical reality.

The BAIF-European Union (EU) Project

An endeavour is now in process to bring all these experiences, together under one development programme, through a BAIF-EU Project, "Transfer of Technologies for Sustainable Development". The Project is supported by the European Union and is being implemented by BAIF under the auspices of the Government of India and NABARD (National Bank for Agriculture and Rural Development). It is in this project that the Jana Utthan Approach has been evolved.

The BAIF-EU Project commenced in April 1996. It is a nine-year Project, being implemented in 5 States and 11 Districts as shown below:

Table 12.1: Geographical Spread of the BAIF-EU Project

Sl. No.	*State*	*Districts*	*Blocks*
1.	Karnataka	Tumkur, Uttar Kannada, Dharwad	Tiptur, Dasankoppa, Kalghatgi
2.	Maharashtra	Ahmednagar, Thane	Akole, Jawhar
3.	Gujarat	Surat, Bharuch	Devgadh, Valia
4.	Rajasthan	Banswara, Udaipur	Ghatol, Jhadol
5.	Uttar Pradesh	Pratapgarh, Gonda	Shivgarh and Gaura, Paraspur

The Project is being implemented across diverse socio-economic/socio-political regions of the country.

Purpose of this Project[2]

Economic Status, health status and women's status of 33,000 disadvantaged poor families improved and stabilized in 11 Districts through a process, that is economically viable, agronomically sustainable, socially equitable and technically replicable.

Up to of December 2001, 217 villages have been covered, and 33,069 family units have been enrolled, thus exceeding the target of 33,000 family units.

The major economic interventions taken up under the Project are improved agriculture, horticulture development, livestock development, and promotion of small enterprises. The Project is also promoting processing units for agriculture produce, and establishing market linkages for products like milk, fruit, pickles, etc.

Another major effort in the project has been the promotion of people's organisations to enable them to take control of the development process in the long run. These organisations can continue to provide essential support services to all the members of the community, while exploring opportunities to take advantage of the programmes launched by the Panchayat Raj Institutions and other donors.

Under the Community Health Programme, it has been ensured that all villages get perennial access to safe drinking water. The Project is also addressing health issues like hygiene and sanitation, malnutrition (severe childhood malnutrition levels have come down to less than 3 per cent across the Project Areas), reproductive and child health issues, provision of primary medical care and control of endemic diseases like Tuberculosis. All these steps have not only improved the quality of life, but also contributed to poverty alleviation by reducing household health expenditure and morbidity.

Women's Self Help Groups promoted by the Project are playing an active role in addressing social evils like spouse harassment, dowry, illicit liquor etc. Micro credit provided by these Self Help Groups has helped many members to come out of the clutches of moneylenders. Women have also been empowered through drudgery reduction by promotion of smokeless stoves and pressure cookers, provision of drinking water, improved flour mills and through specific family/group micro-enterprises.

Evolution of the Jana Utthan (JU) Approach

The Jana Utthan Approach evolved with experience gained in the earlier stages of implementation, through a form of action learning.[3] It evolved through three main stages.

(a) *Technology driven approach*: In the first year, the Project was based around a preconceived set of technical interventions. Each participant family was assigned an intervention, on the basis of target numbers to be achieved per site. It soon became apparent that

- The approach was too inflexible for farmers' needs, and
- The approach would not enable the poorest to improve their economic status. This was partly a result of the technical emphasis in the Project that was geared towards enhancing general productivity, rather than addressing the specific needs of the rural poor families.

(b) *Family based approach:* The Project Team then focused on families' needs. Family level Micro level plans (MLPs) were drawn up, which identified the assets with each family (human, physical, natural, financial, and social), noted the "gap" between the present economic status and the poverty line, and planned for a basket of economic interventions designed to bridge this "gap". This approach was flexible, and effective in targeting the poorest families. However, implementation on a large scale was found to be difficult.

(c) *Community-based Approach*: Partly in response to the increased difficulties in implementation, but more importantly in response to the future sustainability of the development process, People's Organizations (POs) of the Project participants were started. The POs are expected to gradually take over the implementation and monitoring of the MLPs of each family, and to sustain the Jana Utthan Approach after the Project is officially over. The People's Organizations have taken different forms in the different areas, and have reached varying levels of sophistication and maturity. In essence they comprise 3 tiers: operating at hamlet level, village level and cluster Level, with each of the tiers having a different function in the implementation of the Project's activities and control of funds. These POs operate alongside other local POs including women's Self Help Groups and smaller groups like water users' groups, Bal Panchayats, etc. There is also cadre of trained community workers developed by the project who assist in the project delivery. These workers are trained in extension techniques and in specific technical skills, e.g. agriculture, livestock, horticulture, community healthy, gender and empowerment issues etc.

The Jana Utthan Approach took its present shape in the 3rd Year of the Project, and further implementation of the Project's Activities has been in line with the Approach. The essential components of the Approach are:

(a) *Livelihood Systems Component:* This is based around Micro-level Plans for each of the participant families, which considers a family's assets and plans to implement a basket of interventions, according to its capacity. The inputs strengthen the existing assets with a family, and create new ones around them, to bring the family above the poverty line. Inputs include land and livestock based programmes, and small-scale microenterprises in the rural non-farm sector. Each activity is credit as well as subsidy based.

(b) *Area Based and Sectoral Components:* These are designed to complement the Livelihood Systems, by strengthening the infrastructure, environment and social assets of the poor. These include:

- Cattle Breeding Centres,
- Land Development and Watershed Development Programmes,
- Community Health Programmes, which contribute to well-being and quality of life, and
- Women's Development Programmes, which provide a forum for women's self-expression, reduce their drudgery and engage them in more productive activities including livelihoods.

(c) *People's Organizations (POs)*: These have been discussed in the previous paragraphs. An important point to be noted is that these POs fund themselves through people's contributions, received as partial repayments of the project's grant-based livelihoods inputs. The per cent to be repaid is decided by the POs. This has been an important factor in the success of the Project. Since people now have a personal stake; responsibility and ownership of the development process is encouraged, thereby providing a link between livelihoods and wider social initiatives. The PO's funds are presently in Banks, and it is anticipated that the POs will utilize them for credit-based livelihoods support as well as other development initiatives.

(d) *Institutional Linkages:* The Project has been able to establish important linkages with the local Government Institutions, other NGOs and Banks. This has meant that Project activities are not duplicated, but are supplemented. They are also supported, for example by the Health Departments. These linkages have been developed with the involvement of the POs, who in turn have been further empowered. This is essential in the light of the future sustainability of the local development process. The linkages have also assisted the Project and the POs to realise that some of the key elements of the economic and social development chain are already present. It is only the missing links, which need to be provided.

Implementing the Jana Utthan (JU) Approach

(a) The Project started with a Year '0', which enabled the Teams to explore the communities, as well as prepare themselves for the work ahead.

(b) The actual project implementation has been planned for a duration of 8 years, clearly stating the need to provide adequate time for the poor to rise above the poverty line.

(c) *Family Identification Registers*: These are compiled on the basis of a one-page asset based questionnaire for all families in a particular area. It assists the Project Team in identifying the most needy amongst the poor.

(d) *Participatory Situation/Problem Analysis*: The Project Teams conduct frank meetings in all villages, clearly stating the Project Purpose, and obtain the villagers' assistance in identifying the poor and needy families. The very definition of poverty in the local context is arrived at with the help of the people. The various categories of the poor are identified, e.g. W1-W4.

(e) *Entry Point Activities*: These meet the immediate needs of a village. For example, drinking water needs, repairs to a village school, artificial insemination services for cattle etc. They help establish a rapport with the people.

(f) *Preparation of village level Ready Reckoners of economic activities*: These are prepared through Focus Group Discussions and cover the main economic activities undertaken in the village.

(g) *Preparation of Family Level Micro-plans*: Village PO representatives and the poor families together prepare family micro level plans (MLPs), which evolve and are updated every year. Detailed working on MLPs on a small number of families through the first two years, enables identifying of a smaller number of intervention models (sets of interventions) which are then applied on a 'thumb-rule' basis to other families thus enabling rapid up scaling. The village level ready reckoners are utilized to cross-check the present estimate of family economic status prepared with each participant family.

(h) *Credit-cum-Subsidy*: The project works with an indicative, but flexible limit of Rs. 15,000 per family unit. No project input is totally free of cost. Participant Families have to pay in kind or cash, and the latter is deposited in the bank accounts of the people's organizations.

(i) *Training*: All participating families undergo rigorous training in managing the inputs.

(j) *The Role of Community Extension Workers*: A cadre of community extension workers assists the project staff to deliver the livelihoods component, and also the health and women's development components.

(k) *People's Organizations:* These are constituted by the participating families, and support their MLPs by managing input supplies, training, services and monitoring progress.

(l) *Establishing Linkages* with Governments, markets, banks, etc. This is done by project staff along with representatives of the People's Organisations.

(m) *Linkages with the Panchayati Raj Institutions (PRIs)*: The JU approach strengthens grassroots level democracy, and contributes towards better village level plans. This in turn will help strengthen PRIs.

Thus, the poorest in an area, benefit from the implementation of the JU approach in terms of technology transfer, training, organization, infrastructure, marketing, improved social services and conditions and also credit and subsidy-based livelihoods support.

Monitoring Tools

(a) Annual work-plans are the key to the project's implementation in each location and are used as a primary monitoring reference. Variance analysis and other achievement reporting is done through the project's MIS.

(b) The poorest are identified using the ready reckoner of economic activities (described above). This is a useful tool, which is a set of economic projections for all major economic activities in a village. This information is supplemented using data from the family identification register and participatory poverty ranking exercises.

(c) Each family has a micro level plan, specifying the economic interventions planned by the project.

(d) Each intervention is specified in an Intervention Master Plan.

(e) A set of desired characteristics is aimed for in each family and is called "Hallmarks of a Jana Utthan Family".

(f) A baseline information document is prepared for the entire Project Area.

(g) A family passbook/register maintains details of all inputs received by the family. The Register is kept at the project field office while the Passbook is with the family.

(h) A family rating system is under development, which gives the first information about the progress of each individual family.

(i) Progress of the families in coming above the poverty line is monitored through an ongoing participatory impact study on a 5 per cent stratified random sample of participating families, conducted on a yearly basis.

(j) Ongoing participatory impact studies also monitor experiences of the men and women, and their perceptions. These are conducted on a yearly basis, in two villages at a time. These studies are of a qualitative nature, examine progress *versus* the project's logical framework and bring in socio-economic and gender perspectives to the Project's activities.

(k) Technical studies are conducted under each of the Project's components.

(l) Field visits and regional review committee meetings are conducted periodically by project staff and also by NABARD monitoring officers.

Replicating the Jana Utthan (JU) Approach

As mentioned earlier, the Jana Utthan Approach is still in its evolutionary phase. Some of the important learnings thus have been listed here.

(a) A committed inter-sectoral team and team work is essential in implementing such a complex project,

(b) A decentralized management approach emphasizing flexibility and some experimentation is essential to plan for the poor in a given area,

(c) A sufficiently long period of time is needed to implement a Project of this nature, and to allow the poor to make significant progress,

(d) Area based programmes like community health, women's development etc. are essential for the success of poverty alleviation programmes.

The challenges lie in providing technical support and close monitoring of the progress of the families, to assess if they are really coming up above the Poverty Line. This will help policy makers in estimating the resources needed to implement such programmes.

The Project Team feels that the Government of India can replicate the JU Approach. The need is for a committed inter-sectoral team, working for a group of villages.

This approach argues that grassroots people's organizations are necessary, despite active PRIs. They enable development programmes to reach the poorest, and also develop and strengthen the grassroots democracy process. While NGOs like BAIF can play the catalyst role in the process, the government, too, can facilitate the development of people's organizations to implement the approach.

A Comparison with Some Other Approaches

It is important to appreciate that the Jana Utthan (JU) Approach is not only a livelihoods approach but it tries to go further to address social, gender, equity and other developmental aspects as well. The JU Approach compares well with the Swarnajayanti Gram Swarozgar Yojana (SGSY) of the Government of India. However, where the SGSY expects poverty alleviation within a 3-year period, the JU Approach emphasizes a much longer period, especially for the poorest families and communities. three years is the minimum time for a Self-Help Group to mature and assume a developmental role. The JU Approach is also more decentralized and participatory in nature.

The JU Approach also compares well with the DFID Livelihoods Approach[4], and the CARE[5] Approach as shown in Table 12.2 below:

Table 12.2: The JU, DFID and Care Approach

Feature	*The JU Approach*	*DFID Sustainable Livelihoods Approach*	*Care Approach*
Experience	< 5 Years	< 3 Years	> 10 Years
People Centred	Yes	Yes	Yes
Responsive and Participatory	Yes	Yes	Yes
Multi-level	Not at Policy Level Yet	Yes	Yes
Multi-sectoral	Yes	No examples yet	Yes
Partnerships	No Private Sector	Yes	Yes
Sustainable	Too early to say	Too early to say	Yes
Dynamic	Yes	Yes	Yes
Committed to Poverty Eradication	Yes	Yes	Yes
Centred on assets	Yes	Yes	No
Centred on attitudes	Yes	Yes	Yes
Vulnerability Awareness	Yes	Yes	Yes
Political Economy Aspects	No	No	No

Conclusion

The spread and scope of the present BAIF-EU Project, coupled with the development of the Jana Utthan Approach, holds great potential in being able to evolve a flexible, workable and replicable approach to assist poor rural families and communities to improve their own lives sustainably. Clear methods and systems to address development needs are the likely outcomes of this project. The JU Approach appears to be replicable across socio-economic conditions and agro-climatic zones. It strengthens PRIs, and builds up their capacity to eliminate poverty.

The Project so far, has made a significant impact in the field, and has attracted the attention of policy makers, politicians and community leaders. The lessons learnt can be expected to provide opportunities for large-scale adoption and replication at the national level, and the learnings could also be incorporated in the SGSY of the Government of India.

NOTES

1. Government of India (1999) "Guidelines for the Swarnajayanti Gram Swarozgar Yojana", Ministry of Rural Development, New Delhi..

2. The BAIF—EU Project (1996) "Project Logical Framework", BAIF, Pune
3. Action learning is an educational approach where students devise their own study programmes, and carry it out with original research in the field rather than secondary learning in the classrooms. It is an apt description of the manner in which the JU approach has evolved, through practical experience rather than preconceived notions.
4. DFID (1999) "Livelihood Approaches Compared".
5. Scoone, I. (1998) "Sustainable Livelihoods", IDS Working Paper No. 72.

REFERENCES

Government of India (1999). "*Guidelines for the Swarnajayanti Gram* Swarozgar Yojana", Ministry of Rural Development, New Delhi.

The BAIF-EU Project (1996). "Project Logical Framework", BAIF Development Research Foundation, Pune.

"Livelihood Approaches Compared", DFID (1999).

Scoone, I., (1998). "Sustainable Livelihoods" IDS Working Paper No. 72.

13

Milieu Specific Approach to Sustainable Livelihood

NILADRI RANJAN DASH

Introduction

India has registered progress in every sphere since the advent of independence in 1947. Through concerted efforts, the nation has excelled in the fields of science and technology as well as agriculture. Today, India has emerged as the leader not only in South Asia but also in other developing countries. With abundance of physical resources and the world's largest technical manpower, it certainly has promising future prospects. However, spatial and social imbalance in the levels of development impose a dragging influence on its progress. A few regions and segments of the population lag much behind others; most prominently the agriculturally negative, hilly and forested tracts of the country, accommodating major chunk of the country's tribal population.

Since millennia, through age-old social and economic traditions, the tribes had sustained themselves, maintaining a delicate balance with their ecologically fragile milieu. With progress of time, penetration of various non-tribal formations into the tribal territories destabilized the tribal social and economic modes and caused erosion of their traditional economic resource base. Today, these lagging tribal territories stand as formidable barriers on the path of national progress. Efforts made through planned measures, constitutional provisions, privileges and concessions[1] have provided lip service to alleviate the Scheduled Tribes and Scheduled Areas.

The contention is the basic spatial characteristics of the Indian tribes as manifested in their distribution and heterogeneous socio-economic and historical backgrounds have not been taken into consideration, while planning or implementing the development measures. Accounting for 7.95 per cent of the country's population and with a population of 67.75 millions (1991 census), the tribes of India are distributed over most of the states and Union territories. They are far from homogeneous in cultural traits, social organization and modes of living. Their diversity is kaleidoscopic. They vary in size from large groups spread across many states such as the Gonds, Santals and Bhils, each with over two million people to small isolated communities numbering only a couple of hundred. Not all segments of the population of Scheduled Tribes have been equally exposed to the dominant cultures of the regions, in which they are living today. Further, their responses to the processes of acculturation, modernization or initiation into the democratic polity through the electoral process of public mobilization have been far from homogeneous. There is no pan-Indian tribal consciousness. The only count on which they can be treated as one group is that, generally they suffer from the same overriding manifestations of underdevelopment, besides their characteristic clustering and concentration in the agriculturally negative, hilly, forested and plateau areas of the country. An '*a priori*' planning scheme is bound to fail, if implemented uniformly for such a heterogeneous and widely spread group.

It is pertinent, therefore, to devise measures specific to individual and distinctively intelligible tribal milieux of the country. A preliminary attempt is made in the following text to bring out the finer nuances of the milieu specific approach which could generate sustainable livelihoods in the lagging regions as well as rejuvenate their physical resource base. It would not be an exaggeration to suggest that, milieu specific development planning measures would yield equally positive results in non-tribal tribal areas as well.

The Background

Scholars hailing from different disciplinary backgrounds have by now thoroughly assessed the impact of different government initiatives. Although it is difficult to analyze their findings elaborately, the following summary would help in comprehending the adverse effects of developmental planning on the tribal areas.

Natural resources of the tribal areas, particularly forest resources, exploited as a result of interventions by exogenous groups have left a negative impact on the development of tribal areas and have generally proved detrimental to the economic and social development of the tribes (Bhowmic, 1981; Chidiyal, 1982; Parnandes, 1990; Gadgit *et al.*, *1983;* Gupta *et al.*, 1981; Kulkarni, 1994; Shah, 1990; Viegas and Menon, 1986).

Progress of mining and allied activities and the location of mineral based industries in the tribal areas have led to the destruction of tribal social and economic order without creating alternative avenues for their economic rehabilitation (Ahmal, 1985; Das Gupta, 1978; Joshi, 1982; Rothurmund, 1978; Sachidanand, 1985; Sarkar, 1988; Vidyasthi; 1970).

Resettlement of non-tribal population in the forest-rich areas of tribal concentration as a deliberate policy has resulted in the alienation of the tribes from their traditional resource base, i.e. forests, and has accelerated the process of their pauperization (Ahmad, 1985a).

It has also become evident that in every sphere of welfare, be it education [Ahmad, 1978; Premi, 1974], politics [Pathy, 1982] or economy [Bose, 1981; Murdia, 1975; Pathy, 1976; Pathy, 1974] bulk of the tribal population has not benefited from the processes of development in tribal areas. These processes, initiated by consciously designed programmes are alien to the tribes and their social and economic milieu. Expectedly, therefore, the gains have been impounded by the up-coming elites among the tribes.

The tribes are gradually getting alienated from the land, leading to the accumulation of lands in the hands of more articulate elite groups among the tribes or the immigrants (Banerjee, 1981; Das Gupta 1964; Joshi, 1982; Mathew, 1996; Rothurmund, 1978; Sarkar, 1988; Vidyasthi, 1970). Large scale alienation of the tribes from land has also been due to establishment of military set-ups (Economic & Political Weekly, 1994), national parks [Bopat, 1999] and construction of large scale dams[2] etc.

The failure of the programmes has also been noted in the administrative circle. A Government of India document issued by the Ministry of Home Affairs stated that:

"In the name of tribal development we have spent a lot of money during the last several plans, but when accounting was made at the end of the last period of the safeguards to the tribals in the Constitution, it was found that actual benefits trickling down to the tribals have not been consistent with the promise we have made. Expenditure has not resulted in the benefits that should have accrued to this population. Unfortunately, the developmental programmes for tribal areas did not yield the desired results..."(Government of India, 1975).

It further added that more than 75 per cent of total benefits have not reached the tribals and cautioned "we cannot have development at this rate" (*Ibid.*).

Direction of development in the tribal areas indicated lopsided accrual of benefits in favour of the well to do. As the Commissioner of Scheduled Castes and Tribes observed:

"...benefits of various programmes...generally accrues to the well-to-do sections among these communities and the most backward sections among them who actually deserve the benefits thereof do not derive any benefits from these programmes" [Government of India, 1978-79].

Reasons for this failure has been ascribed to the replacement of the colonial model of tribal administration by neo-colonial forms of extraction of natural resources from the tribal regions [Ahmad, 1991]. The State itself has been found to be one of the main beneficiaries of tribal exploitation, and therefore, "it is maintaining the underdevelopment of the tribals both by not attacking the power of the non-tribal rural oligarchy and treating tribal areas as an internal colony" [Jones, 1978].

Recent Experiences as Harbingers

Current research on the problems of tribal area development stresses on empowerment and autonomy of village level bodies to decide the direction and pattern of resource use in their immediate vicinity. Devastation of local resource base in the tribal areas due to large-scale depletion of forests and its consequential adverse effects on soil and water resources as well as on the local economy and culture has given rise to the idea of local self-rule and control of resources. Examples of the idea being practically and successfully implemented are not uncommon.

The story of the Forest Protection Committee (FPC) programme actively supported and developed by the West Bengal Forest Department in the Arabari Range of Medinapore district in 1973 is well-known today (Patil, 1992). Encouraged by the experience of Arabari Range, the Ford Foundation and the West Bengal Forest Department developed a research and training support programme and allowed the FPC experience to extend within and outside West Bengal (*Ibid.*). Following West Bengal, the governments of Karnatak and Maharashtra have tried to employ the FPC type programme in their tribal areas. Beninger (1992) and Gadgil and Guha (1995; 1992) provide several examples from all over India where the tribes manage the village forest and other local resources themselves extremely efficiently without support from governmental agencies.

Wherever people have organized themselves to manage the local resources, the experiences have been positive. Based on the success story of 38 groups set up by the tribal youth in rural areas of Gujarat, Bhatt (1989) feels that the only solution to the problem of underdevelopment is for the people, especially the lower castes, the tribals and the marginalized to organize themselves. It is argued that, to counteract the excesses of globalization, the marginalized have to be trained to organize themselves (Prabhu, 1998; Punalekar, 1997). A lift irrigation cooperative formed by the tribes of Panchmahals district of Gujarat on the initiative of a leading textile company has been producing excellent results (Barik, 1991). Even the conscious segments among the tribes have been demanding local self-rule under the banner of "National Front for Tribal Self-Rule" (Frontier, 1996).

The Joint Forest Management (JFM) programme of 1988, funded by the Ford Foundation, did encourage involvement of local communities. JFM makes local communities partners in protecting and managing state owned forest resources, leading to legitimized role in the control of resources that they have used for generations. These programmes hold great promise for regenerating India's degraded forests and for empowering local communities (Arora, 1994; Burman, 1995; Compbell, 1992).

Through the Panchayats (Extension to the Scheduled Areas) Act, 1996, the Gram Sabhas have also acquired wide ranging powers including customary law, social and religious practices and traditional management practices of community resources, approval of plans, programmes and projects for social and economic development before they are taken up for implementation and many other responsibilities (Mukul, 1997; Palanithurai, 1999).

These new programmes are, of course not completely trouble free. Sustainability and efficiency of JFM is questionable as the existing social structure makes avenues for the upper class and males to dominate (Sunder *et al.*, 1996). Besides, certain tribes are yet to grasp the official modalities of the programme due to inadequate cooperative education. The vested interests of the traders, merchants etc. also come on the way of its implementation (Prabhu, 1998). Although it is mandatory to have women members in the local committees, their involvement is extremely limited. Success of JFM depends to a great extent on the active participation of the female segment of the community (Correa, 1996).

Furthermore, the Forest and Eco-system Act is a deterrent for JFM. A closer look at different sections of the Bill brings out the contradictory intentions in it. On the one hand, it states that the main thrust is to ensure environmental stability, atmospheric equilibrium necessary for all kinds life form, and hence all kinds of economic benefit should be subordinated to this greater and vital need, on the other hand, under Section 38D it permits the "tree growers" or companies to grow trees and fell them according to their needs (Banerjee, 1996; Chhatre, 1996).

The Act also categorizes the reserved forests totally as state controlled. There is no scope for joint management by the forest dwellers and others. In fact, to deny formation of village forests from reserved forests[3] is a retrograde policy on two counts: firstly, the forest dwellers, namely the tribes, have actually been preserving and protecting these forests creating a good eco-balance and enriched biodiversity[4]; and secondly, the earlier Act (i.e. of 1927) provided for the formation of village forests from the reserved forests (Banerjee 1994). It is clear that the historical rights of the forest dwellers are comparatively reduced in the proposed Act. This would impede formation and execution of JFM, providing instead, ample scope to encourage the bureaucracy-traders-anti-socials to rule supreme (*Ibid.*).

The story of Panchayati Raj in the context of certain tribal areas is also not very encouraging. Padhy and Satpathy (1989) have found that the Paraja tribes of Koraput district of Orissa have lost their old political system and have not conceptualized the new Panchayati Raj system. As a result the process of exploitation by politicians, moneylenders, traders and land-grabbers has increased.

Milieu Specific Approach—An Alternative

As is evident from the preceding examples, measures taken at micro-level in different parts of the country have yielded spatially and socially varied results. While certain sections have benefited others have not; certain areas have prospered others have stagnated or even regressed. Basically two issues emerge at this stage.

One, that no planning programme can be considered in geographical isolation, outside the national or the global context. The idea of 'tribal area development' itself has led many to misunderstand the process of development as autonomous and independent in tribal areas and that a completely separate approach is necessary for their development. It is necessary to keep in mind that while tribal areas may provide a unique socio-geographic space, they are nevertheless, part of the entire geographic space[5] within which they must be analyzed.

While there is no denying the fact that the tribal areas have their distinct nuances of development it may not be proper to close one's eyes to the vital fact that the processes of development in the tribal areas are inextricably intertwined with those of development manifesting elsewhere in the national polity as well as with the changes in the global economy. One is filled with dismay to find that the geographers with their broader interdisciplinary training have not approached the issue seriously. Rather some sociologists have ably comprehended and situated the problem in the correct spatial context [Pathy, 1996; Rao, 1996]. Geographers with their grounding in the regional concept and man-nature relationship should be the most suitably placed to handle the problem of development of an area, specially the tribal areas, where man-nature relationship is very delicate and pronounced. Integrating each and every aspect of the physical and the human environment within the spatial framework, the geographers can contribute fruitfully to envisage the sustainable directions of livelihood patterns in the tribal areas of India.

Second, that no regional planning scheme would apply suitably to all areas. Differential response to the '*a priori*' planning schemes experienced in the tribal areas is primarily due to the heterogeneous character of the Indian tribes and the tribal areas. In fact, India is a nation of variety. Variations are manifested in its physical as well as human landscape. The physical landscape encompasses within itself not only all the extremes, i.e. the oldest and the youngest, the hottest and the coldest, the wettest and the driest, but also all varieties of physical features, soil types, vegetation types and climatic types. The human landscape, on the other hand, not only has representations of all religious and ethnic groups and language families, but also has divergent forms of livelihoods developed through millennia old symbiotic relationship with respective regional ecological set-ups.

Albeit, at the aggregate level, these distinct forms tend to merge and give rise to a pan-Indian culture, more popularly known as 'Indian Culture Complex' [Spate *et al.*, 1967]. The diversity of the regional cultures at meso or micro-level rarely poses in opposition or any danger to the pan-Indian unity. Rather, the healthy interplay between them has strengthened the Indian Culture Complex. But, it would be wrong to presume that responses to the spatial processes would be similar over space. Especially, the tribes of India are at various stages of social evolution. Their ecological set-ups differ so also their economic and social formations. Their exposure to exogenous forces have been different so also their social and economic aspirations. Under such a complex situation, milieu specific approach to development seems the only pertinent way out.

Indian scholars, especially social anthropologists, have produced enormous data on various dimensions of Indian tribal lives and livelihoods. Besides, studies conducted by the Anthropological Survey of India provide a vast storehouse of literature on the social characteristics of the Indian tribes. Anthropologists have also studied Indian material culture in its variegated forms—settlement types, housing patterns, tools and implements, rural arts and crafts, clothing and ornaments, and the like. The stupendous work done under the People of India Project led by K. S. Singh is evidence in this regard.

The contention is to identify the micro-social spaces of India's tribal population. Locating each tribal social space within its ecological niche and establishing its local as well external spatial linkages can devise fruitful planning measures. Wherever, information is insufficient or not available at micro-level, further research can help.

For operational convenience, correspondence between the tribal ecological niches with their corresponding watersheds can be identified. The rationale lies in the fact that, the concept of Watershed Development goes well to cope with the environmental concerns, and is an agreeable measure to achieve sustainable development [Fouzdar, 1999].

The Watershed Approach is devised fundamentally to deal with water resources and its scientific management for healthy and sustainable relationship between man and nature. In the context of tribal areas or for that matter any area, a watershed is a human habitat—an ecological milieu of a community, whose social and economic characteristics emerge as superstructure over its infrastructure through centuries of interaction with each other. Traditionally tribal societies function within the broad framework of such milieuts.

What is pertinent is to develop a structural plan of the watershed, which incorporates within its limits a few tribal villages forming a coherent socio-cultural niche. Such correspondence is generally natural, based on which geographers demarcate hill culture, plain culture etc. The structural plan would present the detailed personality of the watershed by developing a synthesis based on all of its perceivable (physical) and imperceptible (social/cultural/ psychological) parameters—a synthesis called 'regional geography' in common academic parlance.

To accomplish the task, watershed mapping of the tribal areas at the smallest order has to be completed at the outset. Invariably it has been seen that at the micro-level, the boundary of the watershed by and large corresponds with that of a few, at times of a single tribal village; wherein the livelihood patterns vary the least. In depth and exhaustive study of such tribal niches based both on the local physical and human resources and the socio-psychological desires of the inhabitants would perhaps help to devise measures for sustainable livelihoods. To elaborate, such a study would require detailed investigation of the following aspects in order to develop a complete database for milieu specific planning.

(i) Complete information on topographical characteristics including geological structure, slope and gradient conditions would be required to plan for the construction of check-dams as well as infrastructure facilities such as school, health centre and other social amenity buildings and structures. Such data would also be of use in analyzing the land use of the area.

(ii) Data on climatic characteristics of the watershed, particularly concerning the amount of average annual rainfall and its seasonality; temperature conditions; and velocity and direction of wind in different seasons, would of course be a prerequisite and helpful in the entire planning process.

(iii) Drainage inventory of the watershed would indicate the type, quantum and seasonality of both the underground water resources as well as the surface runoff. Such information is of immense use for planning agricultural and non-agricultural utilization of water resources. Besides, appropriate sites for construction of check-dams to preserve the surface runoff can also be selected with the help of such information. Erection of such structures at higher slopes in a few tribal villages of Dahod district of Gujarat, with the initiative and technical and financial support of the Foundation for Ecological Security (FES)[6], has recharged the under ground water table of the low lying areas and helped the soil to retain moisture even during the dry season.

(iv) Information on soil types and their characteristics would help in selecting suitable crops of better economic returns traditionally unknown to the tribes. The central Indian tribes generally cultivate coarse variety food grains, such as, maize, bajra, jowar etc. under the dry farming system. After establishing the characteristics of soils and creating sustainable sources of water in the watershed by constructing check dams and/or creating lift irrigation facilities, high yielding variety crops with higher returns can be introduced to the tribes wherever possible.

(v) Much of the forest cover in the tribal areas is lost today. Tracts earmarked as 'forest area' either have sparsely spaced shrub vegetation or transformed into open grasslands. Afforestation of such tracts can be undertaken involving the local population.

Experiences of the Arabari Forest Range in several villages of West Bengal, Maharashtra and Karnataka, suggest creation of village forests and village forest committees for successful afforestation programmes. Direct participation of the concerned population without governmental intervention has yielded better results in many villages of these States[7]. The FES, in all its project areas has been trying to secure tenure of forests and other commons to the traditional/formal village institutions. Experience of the Foundation reveals that, rendering of forests and other commons as government (public) property, and removal of all rights of the community to use and protect them has been the root cause of deforestation. It is essential, therefore, to recognize the viability and importance of the traditional village forums and revive and strengthen them. The Foundation vehemently advocates and has been successful in some of the states, where it has project villages, to secure tenure/custodial rights to village communities over resources like scattered patches of wasteland, stream banks, tank foreshore areas, grazing lands, etc. falling within the boundary of a village on which the community is dependent. It is important that the rights of these communities over the forests, which they have been protecting for a long time, be recognized. Disregard to these historical linkages would alienate the community and could disturb the local forest protection practices. After ensuring rights of village commons in the hands of the community, the approach should be to encourage the villagers to grow species of economic utility for local as well as market use.

The basic database on the physical resources can most easily be prepared with the help of aerial photographs and remotely sensed data. Indian satellites are the most advanced and able to produce imageries of high resolution. The National Remote Sensing Agency (NRSA) is making consistent efforts to further improve upon its capabilities of data collection. Geographers, environmental scientists, botanists, geologists, along with others, have also understood by now, the potential of these techniques in generating data on earth's resources. Such data are proved to be of immense use in geomorphology, urban and regional planning, environmental planning, natural resource management and land evaluation etc. It would not be wrong to expect miraculous results in developing the tribal habitats delineated within small order watersheds through the use of NRSA data. NRSA and other sources of data can help in bringing about the specifications of each and every watershed.

(vi) In-depth study of the human resources; the economic, social and cultural practices and aspirations of the concerned population would also be an important and essential element of the database. For without such information, planning measures would prove utopian. Such studies would not only reveal the ground level reality but also would enable the planning agencies to comprehend the local desires and requirements.

Indian Census provides decennial information on demographic (number of households and number of persons), economic (participation in work), and social characteristics (number of Scheduled Caste and Tribe population and number of literates and educated persons) for the entire country at village level with sex-wise break ups. Information at village level on land use and social amenities are also provided which are fundamental to any planning endeavour.

The Census of India also publishes from time to time, monographs on selected villages of the country. As mentioned earlier, voluminous literature pertaining to various dimensions of tribal life in India, is available today, in the form of research publications of scholars belonging to different disciplinary backgrounds, the government sponsored Tribal Research Institutes, NGOs working in the tribal areas and the Anthropological Survey of India.

(vii) Over and above the secondary level information, every identified tribal milieu must be subjected to detailed field investigation. The objective should be to collate the available information on physical resources in the secondary sources and make an in depth study of the social, economic, cultural and psychological parameters operating in the milieu and their external linkages.

Questionnaire method may help in retrieving quantitative as well as qualitative information on specific aspects. Keeping in view the inherent weakness of questionnaire method to expose the reality from the point of view of the target population, Participatory Rural Appraisal (PRA) technique should be used over and above it. PRA should rather be an essential component of the investigation process. In fact, most of the planning measures in India have been devised from outside, flowing from top to bottom. Consequentially, the efforts have been rendered futile and at times devastating. The development initiatives, thought over in the air-conditioned offices at Delhi and floated as "blanket cover" like measures all over the tribal areas of the country, have invariably been found undesirable for the target population. PRA technique would draw the planners much nearer to reality and enable them to formulate appropriate policies involving the target population itself and in accordance with the ecological and cultural milieu of specific tribal habitats. The technique demands long term association of the developmental agency with the target population in formulating planning measures to create sustainable livelihoods in the area. In the process, it enables the planner to comprehend the desires, aspirations and requirements of the target population and design measures most suitable, based on a holistic approach, i.e. the milieu specific approach. Such an approach carries great potential not only to encourage the female segment of the population to participate in the planning process but also to build a sense of responsibility, confidence and concern in the community as a whole towards the measures taken for their alleviation. With financial, technical and infrastructural support from the external (government and/or non-government) sources, the economic resource base of the tribal areas can be regenerated through afforestation programmes, agricultural extension activities, etc. Quality of the population can also be improved by creation of effective social infrastructure as per the desire and requirement of the concerned population, including health care educational and market facilities.

Conclusion

Emanating from the inherent drawback of "blanket cover" like planned efforts; the tribal areas have been emasculated to generate life-sustaining impulses from within. Planning from the top, without proper comprehension of micro-level circumstantial complexities, has proved extremely undesirable and unsuccessful. Innumerable instances are available suggesting that the highly degraded tribal habitat, bereft of its economic resource base, has adversely affected each and every aspect of tribal life. Local empowerment and participation are of course being proposed as alternatives based on experiments from discrete parts of the country, wherein one or two aspects have been targeted and success achieved. However, sustainable livelihood cannot be expected to base on segmented aspects. An all-pervasive effort has to be made. The proposed milieu specific approach is envisaged with such a perspective, which by its very nature is capable of initiating developmental impulses from within.

NOTES

1. For detailed appraisal see Pathy [45, pp. 165-166].
2. Among different communities that have been affected adversely by government and non-government projects, 75 per cent are stated to be tribes. It is estimated that of the 185 lakh people displaced by various projects between 1951 and 1990, only 52.5 lakh have been resettled. For the remaining 132.5 lakh, resettlement still remains a distant dream [22].
3. For a detailed discussion on the Forest Bill and its impact, see Baviskar [10] and Kulkarni [34].

4. As is observed by Viegas and Menon [59], the "constructive dependence" of the tribes on the forest has been changing to "destructive dependence".
5. Geographic space refers essentially to the earth's surface and the varying attributes of that space. These attributes comprise not only the more easily visible features due either to natural processes or human construction, but also non-visible valuation due to social significance.
6. The Foundation for Ecological Security (FES), formerly known as the National Tree Growers' Cooperative Federation (NTGCF) was started by the National Diary Development Board (NDDB), Anand, Gujarat in 1986 at the request of the National Wasteland Development Board. It works with the rural communities in 10 project areas in 7 states to restore and regenerate degraded land and water resources.
7. See Gadgi and Guha [27 & 28] for more such examples.

REFERENCES

Ahmad, A. (1985). "Identification of Tribal Regions and Planning for Tribal Area Development," UGC Workshop on Research Methods in Regional and Urban Planning (Mimeo) CSRD, JNU, 25 March-13 April, New Delhi.

Ahmad, A. (1985a). "Regional Development Process and Redistribution of Tribal Population in Mid-India", in L. A. Kosinki and K. M. Elahi (eds.) *Population Redistribution and Development in South Asia*, D. Reidel Pub. Co., Dordrecht, 65-78.

Ahmad, A. (1999). *Social Geography*, Rawat Pub. Jaipur and New Delhi, 157-58.

Ahmad, Karuna (1978). "Towards Equalty: Consequences of Protective Discrimination," *Economic and Political Weekly*, XIII, 69-72.

Arora, Dolly (1994). "From State Regulation to People's Participation: Case of Forest Management in India," *Economic and Political Weekly*, XXIX (12), 691-698.

Banerjee, Binay (1994). "No Forest for Forest Dwellers," *Frontier*, XXVII(13), 9-10 and 12.

Banerjee, S. (1981). *Impact of Industrialization on the Tribal Population of Jharia-Ranigunje Coal Field Area*, ASI Calcutta.

Bapat, Jyotsna (1999). "National Parks, Land Alienation and Tribal Livelihoods," *Social Action*, XLIX (3), 305-321.

Barik, B. C. (1991). "Tribal Farmers, Lift Irrigation and Rural Development," *Social Change*, XXI (2), 39-47.

Baviskar, Amita (1994). Fate of the Forest: Conservation and Tribal Rights," *Economic and Political Weekly*, XXIX (38), 2493-2501.

Benninger, Aneeta (1992). "Village Eco-Development: Some Methodological Issues", *Social Change*, XXII (1), 55-60.

Bhatt, Anil (1989). *Development and Social Change: Micro Action by Weaker Sections*, Sage Pub., New Delhi.

Bhowmic, P. K. (1981). "Forestry, Tribes and the Forest Policy in India" in L. P. Vidyarthi (ed), *Tribal Development and its Administration*, Concept, New Delhi, 24-38.

Bose, P. K. (1981). "Stratification Among Tribals in Gujarat," *Economic and Political Weekly,* XVI (6), 191-196.

Burman, J. J. (1995). "Joint Forestry Management (A Case Study of Thakar Tribe of Thakarwadi)," *Vanyajati*, XLII (4), 18-22.

Campbell, J. Y. (1992). "Joint Forest Management in India," *Social Change*, XXII (1), 36-54

Chhatre, A. (1996). "A Socio-Ecological Basis for Natural Resources Management: Forest Bill Debate," *Economic and Political Weekly* , XXXI (18), 1084-1090.

Chidyal, V. C. (1982). "Tribes and Forest" in K. S. Singh (ed), *Economies of the Tribes and their Transformation*, Concept, New Delhi, 133-143.

Correa, M. (1996). "No Role for Women: Karnataka's Joint Forest Management Programme," *Economic and Political Weekly* , XXXI (23), 1382-1383.

Das Gupta, P. K. (1978). *Impact of Industrialization on a Tribe of South Gujarat*, ASI, Calcutta.

Das Gupta, P. K. (1964). "Impact of Industrialization on Tribal Life," *Bulletin of the Anthropological Survey of India*, XIII (1&2), 85-106.

Economic and Political Weekly, (1994). "Netarhat Project: Biggest Ever Tribal Displacement", XXII (18), 1055-1056.

Fernandes, W. (1990). "Forest Policy: A Solution to Tribal Deprivation," *The Indian Journal of Social Work*, LI (1), 35-36.

Fouzdar, D. (1999). "Watershed Development: Concepts and Strategies—Land Water Management Options," *Yojana*, XLII (10), 26-30 & 49.

Frontier (1996). "Tribal Autonomy: Endless Wait," XXVIII (46), 10-13.

Gadgil, M. *et al.* (1983). "Forest Management and Forest Policy in India," *Social Action*, XXXIII (2), 127-155.

Gadgil, M. & R. Guha (1995). *Ecology and Equity: The Use and Abuse of Nature in Contemporary India*, Penguin Books, New Delhi.

Gadgil, M. & R. Guha (1992). *This Fissured Land—An Ecological History of India*, Oxford Univ. Press, Delhi.

Government of India (1978-79). *Report of the Commissioner for Scheduled Castes and Scheduled Tribes*, 26th Report, Part-I, Government of India Press, New Delhi,102.

Government of India (1975). *Tribal Development in the Fifth Plan: Some Basic Policy Papers*, Vol. II, Ministry of Home Affairs, New Delhi, 98.

Gupta, R. *et al.* (1981). *Tribal Unrest and Forest Management in Bihar*, IIM, Ahmedabad.

Jones, S. (1978). "Tribal Underdevelopment in India", *Development and Change*, IX, 41-70.

Joshi, R. S. (1982). "Impact of Industrialization on Tribals," in K. S. Singh (ed.), *op. cit.*, 369-380.

Kulkarni S. (1994). "Proposed Forest Act: An Assessment," *Economic and Political Weekly*, XXIX (30), 1909-1912.

Kulkarni, S. (1983). "Towards A Social Forest Forest Policy," *Economic and Political Weekly*, XVIII (6), 191-196.

Mathew, A. (1996). "Displacement Due to Mining in Jharkhand", *Economic and Political Weekly*, XXXI (24), 1524-1528.

Mukul, (1997). "Tribal Areas: Transition to Self-Governance," *Economic and Political Weekly*, XXXII (18), 928-929.

Murdia, R. (1975). "Land Allotment and Land Alienation: Policies and Programmes for SCs and STs," *Economic and Political Weekly*, X (2), 1204-1218.

Padhy, K. S. *et al.* (1989). *Tribal India*, Indian Book Centre, Delhi.

Palanithurai, G. (1999). *New Panchayati Raj System at Work: An Evaluation*, Concept, New Delhi.

Patil, S. (1992). "Status of Forest Protection Committees," *Social Change*, XXII (I), 32-35.

Pathy, J. (1976). "Political Economy of Jharkhand," *Man in India*, LVI (1), 1-36.

Pathy, J. (1982). "Politics of 'Reservation' in India: Case of Political Agitation in the States of Bihar and Gujarat," *Philosophy and Social Action*, VIII (2), 9-16.

Pathy, J. (1996). "The End of Millennium: Liberalization and Accentuation of the Tribal Problematique," *Indian Journal of Public Administration*, XL (3), 485-489.

Pathy, J. (1984). *Tribal Peasantry: Dynamics of Development*, Inter-India Publications, New Delhi.

Piraburam, B. (1995). "Members' Active Participation in Tribal Cooperative: A Micro-Study of Car-Nicobar Islands," *Vanyajati*, XLIII (3), 23-29.

Prabhu, P. (1998). " Tribal Movements: Resistance to Resurgence," *The Indian Journal of Social Work*, (Sp. Issue: Towards People Centred Development, Part-I), LX (1), 230-255.

Premi, K. K. (1974). "Educational Opportunities for the Scheduled Castes: Role of Protective Discrimination in Equalization," *Economic and Political Weekly*, IX (4), 1902-1910.

Punalekar, S. P. (1997). "On Dialectics of Voluntary Organisation and Social Change," *IASSI Quarterly*, XV (4), 67-90.

Rao, P. V. (1996). "Liberalization and Scheduled Tribes: Emerging Challenges," *Indian Journal of Public Administration*, XLII (3), 490-496.

Rothermund, D. (1978). "The Coalfields—An Enclave in a Backward Region" in D. Rothermund and D. C. Wadhwa (eds.), *Zamindars, Mines and Peasants: Studies in the History of an Indian Coalfield and its Rural Hinterland*, Manohar, New Delhi, 1-20.

Sachidananda, (1985). *Industrialization and Social Organization: A Study of Tribals of Bihar*, Concept, New Delhi.

Sachidananda, (1972). "The Tribal Situation in Bihar" in S. C. Dube (ed.), *Tribal Situation in India*, IIAS, Shimla, 169-184.

Sarkar, A. (1988). "Impact of Industrialization among the Dodhias of Gujarat," *Human Resources*, XXXVII (3), 243-230.

Shah, A. C. (1990). "Creating a Stake in Protecting Forests," *Indian Express*, 20 January, 5.

Spate, O. H. K. *et al.* (1967). *India and Pakistan: A General and Regional Geography*, Methuen, London.

Sunder, N. *et al.* (1996). "Defending the Dalki Forest, Joint Forest Management in Lapanga," *Economic and Political Weekly*, XXXI (45-46), 3021-3025.

Vidyarthy, L. P. (1970). *Socio-Cultural Implications of Industrialization in India: A Case Study of Bihar*, Planning Commission, New Delhi.

Viegas, P. and G. Menon (1986). "Development Patterns, Forest Policy and Exploitation of Tribals" in W. Fernandes (ed.) *Inequality: Its Bases and Search for Solutions*, Indian Social Institute, New Delhi, 57-84.

14

Sustainable Agriculture: Its Goals, Strategies, Concepts—An Analysis

M. ASOKHAN, L. NIRMALA and G. RANGANATHAN

"Sustainable agriculture is not merely a collection of discrete farming practices, but is a vision for the future. Its success ultimately depends upon our willingness to accept and implement a new philosophy of agricultural production."

The story of India's goal to increase production and keep up with population growth, illustrates the wonderful success of the green revolution, as there was wide adoption of high yielding varieties, irrigation and fertilizers. However, they now realize that the future increases must come from less intensive input systems. Sustainable agriculture must not become just another "technological fix" to the problems of agriculture today, but rather it must satisfy the needs of both the present and future generations. "Modern" agriculture is no longer appropriate in an environmentally troubled and energy-poor era. Movement toward a self-sustaining, resource-conserving, energy-efficient, economically viable, and socially acceptable agriculture, i.e., sustainable agriculture, is needed.

Scientists and policy makers around the world are searching for economically sound and environmentally sensitive methods to produce the food, needed for an exploding global population. As we explore socially acceptable ways to bring our human numbers into more realistic balance with physical space and resource limitations, the international community in agriculture is testing alternative farming practices as well as implementing national policies that are designed to assure enough food.

American Society of Agronomy (1990) is of the view that a sustainable agriculture is one that over the long term (a) enhances environmental quality and the resource base, on which agriculture depends, (b) provides for basic human food and fibre needs, (c) is economically viable, and (d) enhances the quality of life for farmers and society as a whole. Reganold *et al.* (1990) opined that, to be sustainable, a farm must produce adequate food of high quality, be environmentally safe, protect the resource base and be profitable. Hailu and Runge-Metzger (1992) suggested that the sustainability of food production systems involves both environmental and socio-economic dimensions.

Many in the conventional agriculture community feel that supporters of organic agriculture don't understand "real" commercial farming. And that sustainable agriculture, too, represents an imposition on farmers by outside forces, environmentalists, ignorant consumers and other zealots, who do not understand and do not appreciate real farming.

"*Sustainable agriculture*" is a loosely defined term for a range of strategies to cope with several agriculturally related problems causing increasing concern in the U.S. and around the world. Notable among these problems are the contamination of the environment by pesticides, plant nutrients, and sediments; loss of soil and degradation of soil quality; vulnerability to shortages of non-renewable resources, such as fossil energy; and most recently, the low farm income resulting from depressed commodity prices in the face of high production costs. Sustainable agriculture is based on several general concepts. They are as follows:

1. Diversity of crop species to enhance the farm's biological and economic stability with techniques such as rotations, relay cropping, and intercropping;
2. Selection of crop varieties and livestock that are well suited to the farm's soil and climate, and that resist pests and diseases;
3. Preference for farm-generated resources over purchased materials, and for locally available off-farm inputs, when required, over those from remote regions;
4. Tightening of nutrient cycles to minimize loss of nutrients off the farm, such as by composting of livestock manures and by rotations using legumes to fix nitrogen;

5. Livestock housed and grazed at low density, with preference for high roughage rations over concentrated feeds for ruminants, and with herd size scaled to the farm's ability to produce feeds and use livestock manure efficiently;
6. Enhancement of the soil's ability to take up applied nutrients for later release as needed by the crop, in contrast to direct uptake by the crop at the time of application;
7. Maintenance of protective cover on the soil throughout the year, through tillage that leaves crop residues on the surface, and through cover crops and living mulches;
8. Rotations that include deep-rooted crops to tap nutrient reserves in lower strata, and that control weeds by alternating between cool season and warm season crops;
9. Use of soluble inorganic fertilizers, if at all, only at a level that the crop can use efficiently, and only to the extent that nutrient deficits cannot be met first by livestock manures and legumes; and
10. Enhancement of conditions for controlling or suppressing weeds, insects, pests, and diseases, with synthetic insecticides and herbicides used, if at all, only as a last resort and only when there is a clear threat to the crop.

Goals of Sustainable Development

Sustainable Development has a fixed set of goals, viz., Basic needs of all human beings must be met adequately, Development processes should be so articulated that ecological balance and environmental purity is least disturbed. All Nations and people must join hands to create world in which the above two goals are optimized.

The alternative strategy to ensure sustainable development in villages is the participatory approach, which needs to be strengthened. Sustainable development is not the business of the government and private companies alone. It includes, Environment-friendly technology, Change in the style of development, Change in our attitude towards nature and Involves political, social and ethical changes.

Evolution and Current Status

Sustainable agriculture's principles and techniques come from many sources, some recent, some going back several decades. It clearly owes an important debt to organic farming, a term that came into use sometime before Second World War. Initially, organic farming emphasized recycling of farm-generated nutrient sources and discouraged bringing in nutrients in the form of livestock feeds and chemically processed fertilizers (Scofield, 1986). Today the term particularly emphasizes avoidance of synthetic pesticides.

Recently, this approach has evolved into a more general concept, variously labelled "sustainable agriculture" or any of several related terms. This evolution has come about as a broader range of people have become interested in the same goals for diverse reasons. The depressed farm economy has caused a shift in emphasis from *maximum production* to *optimal production*, the idea being that a system should be evaluated not simply by how much it produces, but rather by the relative value of what comes out compared to what goes in. This rethinking has given rise to the concept of *low input* agriculture, a reversal of a long-standing trend towards greater production through greater use of purchased inputs.

Sustainable agriculture really denotes a multi-dimensional concept, and it is not surprising that no single view of it has so far gained universal acceptance. It is, therefore, important to keep in mind, many fundamental questions that still need to be discussed, analyzed, and debated before we can know what "sustainable agriculture" really means. Some of these questions may never be answered, but sometimes they aren't even asked, or the answers are asserted rather than demonstrated. Until these questions are discussed explicitly, we cannot know whether we are using the same words to mean different things, whether a particular approach can simultaneously serve the several different goals we would like it to achieve, or even whether there might not be intrinsic contradictions among the various goals that have been thrown together under the single umbrella term "sustainable agriculture". Nor will we know the best economic, institutional, and political environment to allow sustainable agriculture to fulfil its potential.

Fundamental Questions

What are the differences between "sustainable," "alternative," "low input," "ecological," and "regenerative" agriculture?

In the past decade all these terms, plus several less common ones, have come into use as labels for agricultural systems that share basic goals: reduced use of purchased inputs, especially toxic or non-renewable ones; less damaging to the environment; and better protection of water, soil, and wildlife. But both the technical and popular literature is unclear about the exact connotations of these terms.

Strategies for Sustainable Farming Approaches

- Developments of farming approaches that are eco-friendly like natural farming, organic farming, biodynamic agriculture, macrobiotic agriculture etc., are essential.
- Agriculture should be respected as the most essential sector in which more funds are to be pumped in.
- Land has to be considered as an organism, and not be exploited beyond its potential mechanically. This should form the crux of agricultural development.
- Lessons have to be learnt from traditional agricultural systems and efforts have to be made to strengthen the networking of institutions involved in the documentation of these systems and knowledge.
- Massive and committed efforts have to be taken to disseminate and sustain them.
- Models of organic farms and similar approaches have to be developed on large scale, focusing on small and marginal farmers.
- Models have to be developed which are location specific of multiple relationships. Uniformity in the model and complete dependence on few alternatives are to be avoided.

"'Alternative agriculture,' also known as 'sustainable,' 'regenerative,' 'organic,' or 'low-input' agriculture." "Sustainable" has a time dimension and implies the ability to endure indefinitely, perhaps with appropriate evolution:

- "Alternative" describes something that is different from the prevailing, or "conventional" situation. Another implication of the word is that farmers should be able to choose among options, rather than having to follow a single prescription.
- "Low input" refers to reduced use of materials from outside.
- "Ecological" refers to the principles and processes that govern the natural environment.

- "Regenerative" implies the ability to recreate the resources that the system requires.

Conclusion

The main purpose of an extension strategy should be to define the process and the extent of gradual transfer of responsibility from the public to the users. To be meaningful, a strategy needs a well-defined time-frame, which allows each phase to be properly prepared, including the training of both extension personnel and users. Sufficient dialogue with the users to determine the most appropriate means of technology transfer and adequate planning, before moves of extension personnel to new regions to avoid disruption or even discontinuation of services as agents are gradually pulled out.

REFERENCES

Hailu, Zegeys and Runge, Metzger (1992). Caliberating a yardstick for sustainability. *Ceres* 24 (6).36.

Judy Green, (1993). From *Farming Alternatives Newsletter*, Summer.

National Academy Press, 2101. Constitution Avenue, N. W., Washington, DC-20418.

Open Questions in Sustainable Agriculture, William Lockeretz (1988).

Peter Goering, Helena Norberg-Hodge, and John Page, 1993.

Reganold, J. P., R. I. Papendick and J. F. Parr, (1990). *Sustainable Agriculture Scientific American* 258 (1).79.

Research Associate Professor, School of Nutrition, Tufts University, Medford, Sustainable Agriculture: *Why Green Ideas Raise A Red Flag.*

Toward Sustainability: *A Plan for Collaborative Research on Agriculture and Natural Resource Management*, National Research Council, 1991.

15

Policy Issues for Sustainable Agriculture in West Bengal

C. R. PATHAK

Introduction

The agricultural sector in West Bengal has transformed the state from a food deficit to a food surplus one by the end of the twentieth century. Is it the land reform policy adopted by the Government, or technological inputs used for high agricultural productivity, or lack of demand for food stuff because of declining purchasing power of the people which may have resulted in forced surplus? It is true that agricultural production has increased but simultaneously population has also increased and per capita availability of food has not changed much. We may look into the structure of the agricultural sector which is more diversified now than before. Has the structure attained the sustainability level? Are there any gaps in the production system?

The above questions are raised when we find that despite the availability of food stuff, not everyone gets a square meal a day and the level of poverty has not gone down below the national level. No answer could be singled out. Let us examine the relative role of the factors in making the agricultural sector productive in West Bengal, and refer to the public distribution system in reaching the required food to the hungry people.

The Government of West Bengal has announced an agricultural policy in the beginning of the year 2002, which could not be implemented because of difference of opinion expressed by the partners of the Left Front Government. The main controversy is whether the agricultural development programme should be left to the farmers who would adopt such schemes so as to adjust to the condition of market economy or Government should maintain its regulatory and advisory role in designing the agricultural programme. The second revised agricultural policy has tried to remove some of the controversies and recommended greater use of information technology and open up this sector to the market economy by modernizing not only agricultural practices, but diversify by introducing agro-processing and agro-based industries. This programme will reduce the loss of perishable agricultural products and the farmers will get proper price of their products. Government has so far protected the agricultural sector to be self-sustainable and would reduce the subsidies provided to the sector phasewise. This changing role of the Government need to be examined in the context of liberalization process adopted by the country.

Agricultural Situation in West Bengal

West Bengal is a surplus state in agricultural production is a myth rather than a reality. It is true that agricultural production particularly paddy has increased at the average rate of four per cent per annum during the last two decades. The rate is highest among the major food grain producing states. West Bengal is the largest producer of rice in India, which accounts for 15 per cent of the national total, while its population accounts for only eight per cent in 2001. The growth in rice production has come primarily from boro rice (winter crop) with high technological inputs, i.e. irrigation, fertilizer, insecticide and HYV seeds.

Agricultural growth has also come from vegetable production, particularly in potato cultivation with highest per hectare yield (23.8 tonnes per hectare) in India and second only to U.P. in production. West Bengal produced about 17 million tonnes of vegetables on one million hectare of land devoted to vegetable production, of which potato alone accounts for 50 per cent. During 1980-81—1998-99 the area under potato has increased from 115.6 thousand ha to 318 thousand ha—an increase of 175 per cent (however, the land area under potato has declined subsequently to 299.7 thousand ha in 2000-01—a cyclical effect). The production of potato has increased from 1.97 million tonnes to 7.67 million tonnes during 1980-81—2000-01, a four times increase.

The gross irrigated area in the state has increased to 66 per cent in 1998-99. The consumption of fertilizer increased from 35 kg of NPK per ha of gross cropped area in 1980-81 to 129 kg in 1999-2000. The cropping intensity has increased from about 145 per cent in 1980-81 to 175 per cent in 1999-2000. The number of electrified shallow tube wells has increased from 21,000 in 1980-81 to 100,000 in 2000-01.

The total area under cereal crops has increased from 5.17 million. ha in 1980-81 to 6.57 million ha in 1999-2000 (5.92 million ha in 2000-01), an increase of only 18.95 per cent. Boro rice alone accounts for an increase of 3.31 million ha or 325 per cent during 1980-81 to 1999-2000. In production, rice has increased from 7.46 million tonnes to 13.75 million tonnes during 1980-81—1999-2000, an increase of 84 per cent while boro rice production increased from 0.86 million tonnes to 4.47 million tonnes—an increase of 420 per cent (It declined to 4.54 million tonnes in 2000-01).

Agricultural growth in West Bengal has been brought about not so much by increase in land area but by increasing the cropping intensity and yield per hectare. The high productivity in agriculture has been contributed by the technological inputs, not by expansion of land area. When we ask the question of the sustainability of this agricultural growth, we need to examine the input structure. Boro rice and potatoes are the two principal crops, which are cultivated during the dry winter season solely dependent on irrigation, primarily from underground sources. Canals, rivers and tanks or ponds dry out with one or two irrigation, and winter being a rainless season (one or two showers due to westerly disturbances though beneficial to wheat or barley or pulses but not enough for boro rice); cultivation depends entirely on the availability of irrigation water from underground sources. Winter boro rice requires about 12.00 million hectares of water on an average and is entirely dependent on ground water sources through electrically energized/diesel operated shallow tube wells. Boro would not have any problem if the ground water potential could be very large.

To combat the lowering of ground water level, the farmers sink their submersible pump sets to reach the falling levels which further would deplete the water resources. While farmers compete in lowering their pump sets to reach the lower level of ground water, the total inventory of the ground water potential would collapse, and may result in total disaster to boro cultivation. So, in the long run, boro cultivation could hardly be sustainable. Aus rice is uncertain, as it depends on rainwater only. There is a need to re-examine the cropping pattern: one suggestion could be the replacement of high water consuming boro rice to less water demanding wheat and barley and pulses cultivation, as all these crops are grown in winter months only.

The Role of Land Reform in Agricultural Growth

In post-Independence period, many land reform measures have been undertaken, first with the abolition of Zamindari system, then by the Bhoodan and Gramdan movements of Vinoba Bhave in 1951, thereafter by enacting land ceiling acts and the establishment of peoples organizations such as Panchayati Raj and institutional changes. The land reform in West Bengal relating to share cropping (*Bargadar*) by way of giving land to the actual tillers of the land—'a right to cultivate' was initiated in 1972 but got the Presidential assent only in 1977 which the present Government could encash in implementing the act. We shall not discuss here the Bargadar system or nature of rights, the actual tillers would enjoy or what would be the nature of sharing of the produce between the land owner and the tillers. The main features of the system is to empower the actual tillers, the right to cultivate (not owned) to the bargadars. The land registered in the name of the share croppers cannot be evicted easily unless go through the litigation and decided by the court of law.

- It is claimed that more than 1.4 million share croppers have been registered with the public authority earning them security of tenure and a series of new rights in the land and credit market.
- The surplus land over the ceiling was distributed among 42.5 million households.
- Title deeds to house site lands were given to 5,00,000 house-holds.
- 400,000 women got the title deeds of agricultural lands.
- About half of the rural households were directly the beneficiaries of land reforms since 1977.
- Rural agricultural wages have been increased substantially.

The government has remained silent about the amount of agricultural land distributed and what was the quality of such land—whether cultivable or not?

West Bengal is a land of small and medium farmers. It is widely claimed that agricultural growth in West Bengal has been brought about by the small and medium farmers who could invest in agricultural inputs from their own income and also from other sources. It is true that agricultural growth elsewhere has been brought by the feudal land lords (*kulaks*) and large progressive land holders who could cultivate profitably the commercial crops (rice is a commercial cròp in Punjab and Haryana) with capital investments.

In West Bengal, the small and medium farmers could bring prosperity in rice cultivation (trying to produce three types and three times in a year of rice—aus, aman and boro) and vegetables with high technological inputs. Below the belts, the marginal farmers who could ill-afford costly technological inputs have been further marginalized and became agricultural labour in his own land by changing hands to the small and medium farmers and became wage earners. They could earn only if the land owner cultivates the land and earn profit after meeting the labour cost. Of course, the land owner has option to cultivate or not depending upon his assured profit. The landless agricultural labour has no option except to work in other's land. There has been steady increase in their number since land reform and technological input-based cropping pattern introduced.

Although per capita availability of food in West Bengal is not better than all India average, it has hardly increased to 400 gram per capita per day (against the norm of 500 gram per capita per day). The industries in West Bengal have declined and non-agricultural employment has not increased, hence agriculture alone cannot solve the problem of low purchasing power of the people. Statistically it has been shown that the poverty level in West Bengal has come down from 35.66 per cent in 1993-94 to 27.02 per cent in 1999-2000 still out of 80 million people, 22 million are below the poverty line, which is an alarming figure for a small state of West Bengal.

Land reform was a necessary but not a sufficient condition for agricultural growth as the amount, quality and location of the vested land distributed vary from one agro-climatic zones to the other. Most of the holdings were uneconomic in size and hence leaked out to the market. The marginal share croppers could not cultivate because of their incapacity of purchasing costly technological inputs. In many cases reverse share cropping has been taking place, i.e. in reality, small and medium farmers cultivate the share croppers' land in which share cropper is a labour. If the share-croppers, land falls within the irrigated areas, could share with others for irrigation, but if the plots are in the unirrigated zone, cultivation is entirely dependent on the success of monsoon. In the majority cases, the lands remain fallow or sold out to meet the domestic liabilities. There was no restrictions imposed on them in disposing of the allotted land to the farmers through land reform measures. With low purchasing power, the poor people could hardly purchase their food stuff even at the subsidized rate or at the minimum support price, hence distress surplus is being generated.

The Structure of West Bengal Agriculture

Twenty three major crops have been identified in West Bengal. Cereal crops mainly comprise rice, wheat, barley, maize and others. There are six varieties of pulses—gram, tur, mung, masur, khesari and others. Non-food grain crops are of three groups: oil seeds, fibres, and miscellaneous crops. Oilseeds include rapeseed and mustard, sesame, sunflower, and others. Fibre crops include jute, mesta and others; miscellaneous crops include sugarcane, potato, tobacco, and tea.

Cereal crops dominate. Out of 13.45 million ha cultivated in 2000-01 cereals occupy 5.92 million ha, which has increased by 6.15 per cent during 1980-81 to 2000-01. Pulses occupy 6.19 million ha or 46 per cent which has declined by 7.64 per cent during the same period. Hence 90 per cent of land is devoted to food grain production. Among other crops, jute (613 thousand ha), rape seed and mustard (436 thousand ha) and potato (299.7 thousand ha) have sizeable area under them. Area under tea has slowly increased to 109.7 thousand ha in 2000-01.

Cereals and pulses together dominate in production also. Total food grain production reached the highest level of 14.85 million tonnes in 1999-2000 which has declined to 13.81 million tonnes in 2000-01. Decline came from aus and aman rice which are dependent on the success of monsoon. Boro rice could maintain its growth because of extensive irrigation. During drought years, all crops get affected adversely.

Among the non-food grain crops, oil seeds maintained their steady growth both in area coverage and in production. Among the fibre crops, jute maintained its steady growth during 1980-81 to 2000-01 with 7.4 million bales without any increase in area. Area under potato increased up to a maximum of 318 thousand ha in 1998-99 and declined to 299 thousand ha in 2000-01 without any decline in production. The maximum production of 7.6 million tonnes was obtained in 2000-01. Sugarcane area inclined slightly from 22.9 thousand ha to 21.6 thousand ha in 1999-2000 and its production declined from 2.0 million tonnes to 1.46 million tonnes during 1998-1999 to 2000-01. Tea experienced slow rate of increase in area but stagnancy in production.

The Government of West Bengal is planning to achieve an annual growth rate of 4 per cent per annum during the Tenth Plan period. To realize this growth rate of 4 per cent in agricultural sector, Animal husbandry, fisheries, vegetables and horticulture, have to grow at the rate of 8 to 10 per cent per annum during the Tenth Plan. There is a structural gap in West Bengal agriculture. The sector has to be diversified. Such diversification in agriculture would be able to support the agro-based and agro-oriented industries and shall strengthen the food security and generate employment and income.

Towards an Agricultural Policy in West Bengal

To make agricultural development in West Bengal, a sustainable one, there has to be a sound policy framework. The development in agriculture sector alone would not be able to re-vitalize the West Bengal economy at the background of continuous declining secondary manufacturing sector since 1980s. Without energizing the secondary sector the primary sector activities cannot grow, as both the sectors are functionally interdependent what to speak of sustainability.

The Government of West Bengal had appointed a foreign firm (Mackenzie) in 2001 to prepare draft report on agricultural policy framework. The firm submitted its report on October 17, 2001, on which the Government of West Bengal deliberated and issued a policy in February 2002. This policy was discussed but could not be adopted since there were difference of opinion among the members of the Ministry. Accordingly, the Government issued the revised policy. The basic objective of the policy is to make the agriculture sustainable by raising it from production of food to agri-business, by diversifying its structure with the application of information technology, with more investment and creating demand for agricultural products.

This new policy was discussed in the academia and in the media at large. We shall discuss the salient features of the policy. The International Consultancy Firm known as Mackenzie was appointed in October 2001 to critically examine the development scenario of the agricultural sector and to suggest some policy issues to make it sustainable. The firm put forward its recommendations and suggested certain reform measures for raising the agricultural sector from subsistence level to agri-business level. If the reform measures are adopted, West Bengal could be the granary of India by 2010. It suggested greater use of information technology in this sector for higher productivity, and diversify the sector by introducing agro-based and agro-oriented industries and make it export-oriented or in short agri-business.

The new agricultural policy will have four interrelated objectives:

1. Sustain the achievements of the state with regard to rice production that will maintain self-sufficiency in food and food security,
2. Increase yield in rice production shall release a sizeable amount of land for crop diversification particularly for the production of the oil seeds, pulses, fruits, flowers, vegetables and other food crops,
3. Protect the bio-diversity and develop agriculture and related activities and in general, plan land use in an ecologically sustainable way, and
4. Ensure employment generation in the agricultural sector, increase income and living standard of the working people in the countryside. Agricultural policy need to be integrated with the rural development having universal provision of basic services to the countryside.

These policies aimed at 4 per cent annual growth in farm sector and for that 8 per cent growth has to be achieved in horticulture, animal husbandry and fisheries. These policies have been commented upon by the media and the academia and even by the partners of the Left Front Government.

We shall examine the policies recommended and find out any alternatives that could be considered in modifying this market-driven reform:

- Land reform in West Bengal giving land to the tillers has motivated the farmers to produce more. But high cost of technological inputs has increased the cost of production while the demand has fallen. The farmers have to sell their products below the support price and has dampened the incentives of production. The situation is paradoxical.

- The Government has put more emphasis on the production side but not on the marketing and distribution side.
- Because of the decline of the manufacturing sector during the last quarter of the century, (agricultural sector could not pay even the minimum wages to the agricultural workers what to speak of generating new employment), has adverse effect on the purchasing power of the people. The decline in the purchasing power may be one of the reasons for agricultural surplus.
- It is no doubt that the agricultural sector need to be diversified not only in cropping pattern but also by making it a commercial venture by introducing agro-processing and agro-oriented industries.
- Investment has to be stepped up in the food processing industries, and processed food may be marketed with some subsidies like adopting sales tax reform. How far the Government could be able to protect this sector in this age of globalization and liberalization from market economy is an open question.

Table 15.1: Availability of Food Grains in West Bengal

	1980-81	*1990-91*	*2000-01*
Population in million (1981, 1991, 2001)	54.58	68.08	80.20
Food grains in 000 tonnes			
Cereals G	8043.2	11076.7	13595.7
Cereals N	7037.8	9692.2	11896.3
Pulses G	238.2	193.4	219.5
Pulses N	208.5	169.2	192.5
Total food grain G	8281.4	11270.1	13815.2
Total food grain N per capita availability per day/gram	7246.3	9861.4	12088.8
Cereals (N)	353.2	390.0	406.4
Pulses (N)	10.4	6.7	6.5
Total (N)	363.6	396.7	412.9

Note: For per capita net availability, total gross production has been deflated by 12.5 per cent for seed, storage loss and others.

Source: Economic Survey of WB 2001.

Table 15.2: Growth of Population and Food Production in West Bengal in per cent

	1981-1991	*1991-2001*
Population	24.73	17.84
Cereals	37.71	22.74
Pulses	-18.84	13.77
Total	36.08	22.58

Source: Economic Review, Govt. of WB, 2001. (Tables 5.5 and 5.6)

Table 15.3: Production of Rice and Wheat in WB

Area	*1980-81*	*1990-91*
Aus Area	615.1	610.3
Production	576.4	906.3
Yield	937	1485
Aman Area	4214.6	4306.5
Production	6024.0	6865.8
Yield	1429	1594
Boro Area	346.5	896.1
Production	865.2	2664.4
Yield	2497	2973
Rice Total		
Area	5176.2	5812.9
Production	7465.5	10436.5
Yield	1442	1795
Wheat total		
Area	283.0	269.1
Production	473.2	530.7
Yield	1672	1992

Source: Same as in Table 15.1.

Table 15.4: Per cent Increase in Area, Production and Yield of Cereal Crops in West Bengal

Item	*1980-81 to 1990-91*			*1990-91 to 2000-01*		
	Area	*Production*	*Yield*	*Area*	*Production*	*Yield*
Aus Rice	-0.78	57.23	58.48	-35.44	-24.54	16.83
Aman Rice	2.18	13.19	9.45	-15.49	4.91	24.15
Boro Rice	158.61	207.95	19.06	56.43	70.44	8.9
Total	12.30	39.79	24.48	-06.50	19.08	27.35
Wheat	-4.91	12.04	19.19	58.30	99.66	24.75

Source: Same as in Table 15.1.

REFERENCES

Chattopadhyay, Suman (2002). From Agriculture To Agri-business and Information Technology-to be equal to Malaysia & Singapore, A. B. *Patrika,* October 30 and 31 (Bengali Daily), Kolkata

Dasgupta, S. (2002). One Government and Two Agricultural Policy—a few questions, A. B. *Patrika,* (Bengali Daily), Kolkata.

Draft Agricultural Policy of West Bengal (2001). *McKinsey Report*, Kolkata.

Economic Review, 2001-2002: Statistical Appendix, Government of West Bengal, Kolkata.

16

Agro and Rural Industries: Development Strategies

CHRISTY FERNANDEZ

The Indian economy is basically agrarian. The agriculture sector continues to be the mainstay of rural employment and principal source of income. But its ability to absorb the ever increasing workforce is limited, for various reasons, particularly due to the labour saving practices progressively being used in this sector. The organized industrial sector also has limitations in absorbing the incremental labour force, and of late it has been witnessing a phenomenon of jobless growth. In this context the rural non-farm sector (RNF) comprising secondary and tertiary activities gains considerable importance as a potential avenue for employment and income generation.

Rural Industry—Backbone for Rural Employment and Income Generation

The Rural Industry (RI) in India is a diverse sector offering a significant contribution in terms of employment, productivity, earnings and welfare. It encompasses both traditional and modern enterprises, mainly utilizing rural resources and rural skills. Small scale units, power loom units, etc. are classified as modern small scale industries. Hand looms, Khadi and Village Industries, handicrafts, sericulture, coir etc. are considered traditional industries. They are generally based on resources like forest, crops, leather, mineral, polymers, chemicals, engineering, non-engineering and services. There are about 95 identified categories of rural industries in the country. They benefit the rural masses, especially those belonging to the marginalized and underprivileged sectors of the society who are not exposed to modern knowledge and skills which are prime requisites in an open economy. The Agro-Rural Industries (ARI) have to be treated as a specialized group of industries based on agricultural resources. Although they have several characteristics which are common to the SSI sector, they are required to be dealt with separately. Hence this paper focuses more on the agrarian-based artisan sector involving tiny, small and medium manufacturing enterprises, primarily located in the rural areas.

Uninterrupted supply of agro based raw-material is a fundamental necessity for sustenance of Agro-Rural Industries. The forests, grasslands and croplands are the major sources for it. Protecting this biological base, forest, grass and croplands is crucial for the sustainability of this sector. With the increase in population, large areas of forests are cleared for human settlements, roads, parking slots, shopping malls, etc. Air pollution, acid rain, overgrazing, unscientific logging and poor farming practices are causing severe damage to the ecosystem. The excessive demand on the biological systems affect their carrying capacity, leading to their degeneration. Deforestation is causing serious threat to the global ecosystem. The earth's forests are shrinking by an estimated 17 million hectares per annum. Protecting the forests is key to the sustainable development of wood-based industries, which is one of the most important categories of rural industries. Efficient use of wood, reducing waste and recycling of paper and other wood products are also methods to save the forests. Aforestation in additional areas, would lead to more trees preventing soil erosion, absorbing more CO_2, slowing the pace of global warming and more wood for industries. The grasslands too have a significant role, in providing raw materials for rural industries. About 24 per cent of the earth's surface is devoted to support livestock, which provide industrial raw materials like leather, wool and tallow, apart from milk and milk products.

Impact of Liberalisation, Privatization and Globalisation

The era of libralisation, privatization and globalisation has brought in its wake a variety of new challenges hitherto not experienced by the ARI sector. With the opening up of market under WTO, and withdrawal of QRs, they are thrown into the vertex of severe competition from within and without, that too in the face of dwindling state patronage and protection. The fast changing consumer preferences and widening choices have made the markets more demanding. The survival of the ARI depends on its ability to adapt quickly to these changes and compete successfully in the new environment. The ARIs, especially the traditional ones are beset with a host of problems relating to decentralised production, like irregular supply and high cost of raw materials, low value addition, poor marketability of products, poor accessibility to markets, poor market orientation, poor quality, inadequate market intelligence, insufficient R&D support for product development and diversification, etc. In this background, a new strategy is required to be evolved to counter the challenges of emerging market economy, particularly with the objective of arresting the declining trend of ARIs. Under the mounting pressure of competition, some highly vulnerable categories of ARIs, especially those in the traditional sectors are perforce bound to decline. It is here that the state intervention in the form of policy support and promotional efforts become necessary. It should pave the way for sustaining ARIs in the transitional stage.

Impact of WTO on ARIs

The WTO Agreements have a direct bearing on the ARI. Competition in the domestic market from goods imported at lower tariffs will be more severe. In the export market, they will face competition from other developing countries and non-trade barriers like environmental standards, occupational and health safety, working condition, labour standards, labour welfare measures, technical barriers like sanitary and phytosanitary measures etc. The ARIs are yet to gain the requisite expertise in these matters. Appropriate capacity building and skill development measures are urgently warranted along with necessary policy interventions to facilitate them to handle these issues successfully. The specialized organisations both at the centre and the states have a significant role to play in building up capacities of the stakeholders. Besides, these organisations have to keep close watch on the movement in the market to see whether any safeguard measures are required to be taken to prevent dumping irrational use of non-tariff barriers, etc. Adopting the role of facilitators, they may help in getting value added market information, which the ARIs on their own will not be able to generate. Although there are several safeguards available under WTO, to obtain due protection under them is a Herculean task. The ARIs would not have the expertise and wherewithal to handle such cases. The government and its specialized organisations must come forward to extend necessary assistance in this regard.

Marketing of products is often an uphill task for the ARIs with severe competition and gradual withdrawal of state protection that they have been enjoying so far. The ARIs cater to both domestic and export markets. The Indian domestic market is expanding and the pent up domestic demand is getting unleashed. For the ARIs, which produce low value added commodities, the domestic market is more lucrative than the export market. Whether it is for domestic or export purposes, the ARIs have to develop the requisite market orientation and produce marketable products at competitive prices. The biggest challenge before the ARI managers in the new era of open market, will be to keep the quality of their products and services high, and their costs low.

An area often neglected in the context of WTO is patenting of products and technologies developed by ARIs. Great risk is involved in not doing so. Agencies like NRDC should come forward and guide ARIs to identify their inventions and get them patented, free of cost.

Modernisation—Key to Sustainability

Modernisation is the key to sustainability of the ARIs. Modernisation of ARIs should aim at use of appropriate technology, conversion of resources and adopt eco-friendly strategies. Obsolescence in technology, design and near absence of product innovation are their major handicaps. For this, R&D efforts have to be geared up. Academy-industry interaction in product/process technology, innovation and design development should be encouraged. Collaborative R&D projects should have a market-orientation, where the customer preferences and taste are brought to the R&D lab for development of new products required by the market.

In the emerging scenario, production efficiency involving lower cost and better quality of goods and services alone can help ARIs to meet with the competition. Managerial inputs hitherto neglected by them have to be adopted to improve cost efficiency, better marketing margins, product designing, information dissemination, logistics etc.

Modernisation and technology upgradation while bringing about positive changes may also give birth to a set of new challenges. Displacement of workers and death of non-viable market unfriendly tiny units are an expected natural phenomenon of technology upgradation in the traditional sector. But it is not the end of the world for the affected parties, as new opportunities like subcontracting and ancillarisation are coming up simultaneously. The ARIs have to get familiarized with this global trend of subcontracting and franchising arrangements. These new avenues bring with them new issues, for which fresh policy initiatives are imperatives. The ARIs have the resilience and capacity to modernize and adjust with them, provided right inputs and supports are offered. But at the pace at which they would accept the changeover to modern technology and systems especially, modern management practices may be very slow. While the ARIs are prompted to adopt modern technologies, they should be encouraged to accept ecologically sustainable production strategies like energy conservation, waste minimization, recycling of materials such as paper, glass, metal etc. Poor working conditions cause occupational health hazards. At least some categories of ARIs cause pollution due to their production technologies, inefficient equipments, poor maintenance and inability to undertake treatment of effluents emitted by them. Creating awareness about hazards of pollution, promotion for cleaner technologies, production of eco-friendly goods and appropriate support mechanisms should form part of the development strategy for ARIs.

Cluster Development—A Practical Approach

A cluster approach has been generally accepted as a practical model for the development of ARIs, especially for the traditional industries. The National Programme of Rural Industrialisation (NPRI) has endorsed it as a key policy instrument. Under the NPRI, it has been proposed to set up 100 rural clusters every year to give a boost to rural industrialization. They are expected to benefit rural artisans and unemployed youths and also reduce rural urban disparities. There are several proven approaches to cluster development involving stakeholders, Self Help Groups, promotional agencies, NGOs, industry and trade associations etc. The cluster development approach would make it comparatively easy to provide forward-backward linkages and build necessary physical infrastructure. They can resort to a more ecologically sustainable course of action, by setting up common facility centres, especially common effluent treatment plants etc. It is an established fact that peoples participation is integral to any development strategy. As such a development strategy for ARIs through cluster approach with the active involvement of Panchayati Raj institutions may be an effective alternative. A vital integration of this nature is perceived to have a lasting impact on the development of ARIs in the years to come.

Self Help Groups and Consortia—An Effective Alternative

Provision of timely and adequate credit has always been a serious hurdle to the ARIs. There are often reports that the formal banking and financing institutions have not been (generally) helpful to the micro eneterprises. Although several institutional arrangements exist for providing credit and to ensure timely payment to small and micro enterprises the ARIs are still finding it difficult to raise adequate resources mainly working capital. An attempt is being made to fill this gap to at least some extent through the SHGs and NGOs. The cooperative institutions of yesteryears have not been able to rise to the occasion as channels for raw material supply, production and marketing. Recognising the importance of the SHGs and Consortia of rural micro enterprises special package of services may be introduced, as they are effective instruments of rural poverty alleviation and empowerment, especially of women workers. Given the right support, they can grow into a nationwide network for micro-financing, production and marketing of goods and services.

A National Policy for ARI

First and foremost efforts should be to identify the policy gaps prevailing in the existing scenario, to meet with the emerging demands. The specific characteristics of the ARIs are different from that of other subsectors in the SSI sector, although there are several common features. Of late, the SSI sector has been getting special attention from the Government, by way of special policy packages which of course are desirable. But the ARI sector is yet to get its due share of policy support, as a sector which offers more employment opportunities than the SSIs in India. What is good for the SSI sector need not necessarily be adequate enough for the more vulnerable ARI sector, dominated by tiny, traditional artisan units. The policy prescriptions have to be more specific to the requirements of the ARIs. Therefore, a countrywide mapping of the ARIs with the specific characteristics of each of the categories and areas that it belongs to has to be undertaken in close association with the State/UT Governments. Based on the data thus generated, a **National Policy for ARIs** has to be formulated and implemented through a time bound Action Plan duly accompanied by earmarked plan allocation, both at the national and State/UT levels. In spite of its significant contribution to the national economy, the share of ARIs has been going down over the years in the overall plan outlays. The share ought to be commensurate with the growth of this sector—at least 2 per cent of the total plan outlay as recommended by the study group appointed by the Planning Commission. The policy statement should essentially reflect the socio-economic concerns of the ARIs and its stakeholders. It should spell out the development and promotional strategies for each category of ARIs and the role of existing specialized organisations like KVIC, KVIBs, Coir Board, SIDO, SISI, Handloom and Handicrafts Development Commissionerates, SIDBI, NSIC etc. The impact of various ongoing schemes under various Ministries and Departments, both at the Centre and the State in relation to ARIs is required to be assessed. The complementarity of these schemes need to be identified for greater synergy and duplication, if any, should be scrupulously avoided. An Inter-Ministerial Task Force under the Union Ministry of ARI or Planning Commission may be entrusted with this responsibility of review of schemes. The Ministry of ARI may be made the nodal agency for coordinating and monitoring the implementation of the National Policy on ARI.

Conclusion

The ARIs, thus, endowed with the potential to be the engine of growth for Indian economy need a more focused attention both at macro and micro levels. While formulating a development strategy for them an orientation towards ecologically sound principles, should find its legitimate place since it is important for their sustainability as the sector of the millennium for the rural India.

17

Maximisation for Interactions Among Micro and Macroflora for Sustaining Agroforestry in Alkaline Soils

SATYAWATI SHARMA, SUMAN KASHYAP
and PADMA VASUDEVAN

Introduction

Alkaline soils are the sick soils suffering from sodicity/alkalinity (excessive exchangeable sodium content), and are generally poor in fertility with low available nitrogen, phosphorus, potassium, manganese, calcium and copper nutrients. The low fertility, high osmotic pressure, reduced availability of nutrients, poor air and water movement and low microbial activity are the important problems which have a direct bearing on the nutrition of plants on salt-affected soils. With anti-plant growth properties, these soils have many adverse effects, e.g causing low yields of crops or crop failure in extreme cases, limiting the choice of crops, rendering the quality of plants poor, causing excessive run off and flood owing to poor infiltration resulting damage to crops in the adjoining area (Murty, 1987). According to the Bureau of soil survey and land use planning, an area of 10.1 m ha is afflicted by this problem in India (Yadav *et al.*, 1990). Of this the state of Haryana has 0.5 m ha of wastelands mainly affected with salt related problems (Murty, 1987). But the increasing demands of the expanding population for food and energy necessitate the use of these soils by exploiting some appropriate systems like agroforestry in alkalinity affected areas.

Agroforestry is a sustainable land management system, combining agri or horticultural crops with tree species with maximum desirable root microbial interactions with the principle of sustained biomass yield and soil conservation and would have the great potential to optimize production and economic return per unit area. In any mixed cropping system, suitable combination of plant species and their rhizospheric microflora is necessary, as there are vast communities of microbes in soil interacting with each other as well as with plant roots. The efforts should be made to promote the native beneficial ones by using the optimal combinations of selected microbes (Janardhanan *et al.*, 1994).

Certain soil properties, i.e. moisture holding capacity, *p*H, texture, organic matter etc. favours the establishment and survival of certain micro-organisms. The alkalinity in the soil may also promote or decrease the activities of certain beneficial microorganisms. Hence, it was proposed to inoculate and culture the beneficial microbes (AM fungi, *Azotobacter, Rhizobium* and *Azospirillum*) under alkalinity at different *p*H in *Trigonella foenum graecum* and *Ricinus communis*. Further, the efficiency of these bioinoculants affecting plant growth of *Morus alba* and *Populus deltoides* under alkaline conditions was studied. The research work was mainly focused on promoting interactions of native micro and macroflora to develop sustainable agroforestry models for alkaline soils in Haryana.

MATERIALS AND METHODS

1. Culturing of AM Fungi under Alkaline Conditions

The native strains of AM fungi with four species of *Glomus* (*G. mossae, G. microcarpum, G. macrocarpum, G. intraradices*) and two species of *Gigaspora* (*G. margarita* and *G. heterogama*) were used for culturing in *T. foenum graecum* and *R. communis*. The multiplication of AM fungal spores was done by pot culturing method. For each plant specie, twenty replicates were taken. The seeds of both plants were sown in sterilized (by dipping in 1% formalin for 24 hours) earthen pots (size 25 cm diameter × 30 cm depth) filled with sterilized soil: vermicompost (3:1) with 4 kg/pot. Alkalinity at different *p*H values (9.0, 10.5 and 12.0) was induced by adding different concentrations of Na_2CO_3 to soil of Micromodel, IIT, Delhi. The isolated mycorrhizal spores were surface sterilized by chloramines T (2%, w/v) and 200 ppm streptomycin and then repeatedly washed thoroughly in sterile distilled water. Fifty surface sterilized spores were inoculated per pot. To see the effect of Hoagland nutrients solution (Hoagland and Arnon, 1950) on culturing of AM fungi under alkaline conditions, it was applied weekly (250 ml/pot) to the pots with *p*H 9.0. Optimum watering was given regularly avoiding any leaching of salts. Data of both plant species pertaining to germination per cent, survival per cent, AM infection per cent (Phillips and Hayman, 1970), AM spore count/100 g soil (Gerdemann and Nicolson, 1963) and fresh and dry weights was recorded.

In another set of experiments, culturing of AM fungi on five different plant species, namely *T. foenum graecum, R. communis, Zea mays, Sorghum vulgare* and *Penniseteum typhoideum* in village soil, (*p*H 9.0), was done to prepare a rich inocula of village native AM fungi for further use in rural area. The AM spores were isolated from Farrukhnagar soil (*p*H 9.0) and identified as G. *mossae*, G. *microcarpum*, G. *macrocarpum*, and *Gigaspora margarita*. These were inoculated in a similar manner as mentioned above. The Hoagland nutrients solution was applied in half of pots to make comparisons.

2. Effect of Bioinoculants on Plants' Growth under Alkaline Conditions

A study was conducted to observe the effect of different *p*H on the growth of two plant species namely *Morus alba* and *Populus. Deltoides* with different bioinoculants. The main objective was to evaluate the effect of different bioinoculants on growth parameters of plants under alkaline conditions, and to produce saplings to be transplanted in village to form suitable agroforestry models. A nursery of above selected tree species with different bioinoculants (AM fungi, *Azotobacter* and *Azospirillum*) was raised in polythene bags at Micromodel, IIT, Delhi during February to March 2002. The soil of Micromodel analyzed chemically was found to be normal with EC = 0.118mmhos/cm, *p*H = 7.50, N = 350kg/ha, P = 18.0kg/ha, K=270kg/ha. The cuttings (15 cm length × 1.5-2.0 cm diameter and with 2-3 axillary buds) of each popular (*Populus deltoides*) and mulberry (*Morus alba*) were taken. These were treated with carrier based inoculum slurry of native bioinoculants (*Azotobacter* and *Azospirillum*). For AM fungi treatment, 100 g soil inocula (with ~ 200 spores/100 g soil) was added per polythene bag. Twenty replicates were kept per treatment and cuttings were planted in polythene bags having Micromodel soil (*p*H 7.5) and village soil (*p*H 9.0). In all the experiments, vermicompost (soil: vermicompost = 3:1) was utilized as organic manure (Tables 17.6-17.9).

Statistical Analysis

The data was analysed by computer using the SPSS for windows 9.0 package. Categorical data was compared using Chi-square analysis and Fisher's exact test when indicated (expected frequency of less than 5 in any cell). ANOVA was applied on quantitative variables with multiple groups, followed by Scheffe's multiple comparison test. Quantitative variables with normal distribution and equal variance were compared by two-tailed t-test. The Mann-Whitney U non-parmeteric test was used for non-normal data.

RESULTS AND DISCUSSION

1. Culturing of AM Fungi under Alkaline Conditions

Data pertaining to AM culturing on *T. foenum graecum* and *R. communis* under high *p*H and effect of Hoagland solution (at *p*H=9.0) are depicted in Tables 17.1 to 17.4. It was noted that AM infection percentage did not change with increase in alkalinity (60% at *p*H 9.0 and 10.5, 59% at *p*H 12.0, 60% at *p*H 7.5) in *T. foenum graecum*. No significant difference was observed with respect to AM spore count/100 g soil, AM infection percentage and number of nodules/plant at different alkaline levels (Table 17.1). Hundred per cent germination and survival percentage was noted in all the treatments during four months period. Although with increase of *p*H from 7.5 to 12.0, fresh and dry weight of plants decreased. The survival of these plants at high *p*H with good AM infection and sporulation indicated the efficiency of AM fungi under alkaline stress. Our results related to potential of AM fungi under stress condition are also in accordance with studies of Pffeifer and Bloss, 1988 and Hirrel and Gerdemann, 1980. AM plants have been shown to have greater tolerance to stress, to saline soils and to adverse soil *p*H than non-mycorrhizal plants (Mosse *et al.*, 1981). It was observed *in situ* and *in vitro* by Dixon *et al.* (1994) that introduction of mycorrhizal fungi to unfavourable saline conditions may improve early plant survival and growth. The AM isolates were found to be effective in enhancing growth of plants on both alkaline and acid soils showing wide adaptability to broad soil *p*H conditions (Clark and Zeto, 1996). AM fungi have been found to influence the composition of amino acids and carbohydrates of the host plant. It has been reported that in many cases, the increased tolerance to salt, afforded by mycorrhiza, is probably due to improved P nutrition (Jindal *et al.*, 1955).

Table 17.1: Culturing of AM Fungi on *T. Foenum Graecum* with Na_2CO_3 (four months data, December to March, 2002)

Treatments pH	*Germination (%)*	*Survival (%)*	*AM infection (%)*	*AM spore count/100g soil*	*No. of nodules/ plant*	*Fresh weight (g)*		*Dry weight (g)*	
						Shoot	*Root*	*Shoot*	*Root*
9.0	100	100	60	83 ± 5.0	7 ± 2.0	7.60 ± 4.5 ^{a}	0.487 ± 0.2 ^{a}	2.85 ± 1.0 ^{a}	0.250 ± 0.1
10.5	100	100	60	83 ± 6.0	7 ± 3.0	7.54 ± 5.2 ^{a}	0.475 ± 0.1 ^{a}	2.82 ± 0.9 ^{a}	0.239 ± 0.2
12.0	100	100	59	82 ± 5.0	6 ± 3.0	6.22 ± 1.2 ^{b}	0.416 ± 0.1 ^{b}	2.6 ± 0.7 ^{b}	0.190 ± 0.1
7.5 (normal)	100	100	60	82 ± 4.0	7 ± 2.0	7.96 ± 4.7 ^{a}	0.493 ± 0.2 ^{a}	2.95 ± 0.5 ^{a}	0.262 ± 0.1

Table 17.2: Effect of Hoagland Solution on Culturing of AM Fungi in *T. Foenum Graecum* with Na_2CO_3 ($pH = 9.0$) (four months data, November to February, 2002)

Treatments	*Germination (%)*	*Survival (%)*	*AM infection (%)*	*AM spore count/100g soil*	*No. of nodules/ plant*	*Fresh weight (g)*		*Dry weight (g)*	
						Shoot	*Root*	*Shoot*	*Root*
+S +H	99	90	72	120 ± 7.0 ^{a}	15 ± 3.0	8.4 ± 0.7	0.579 ± 0.1 ^{a}	3.22 ± 1.2 ^{a}	0.224 ± 0.2 ^{a}
+S -H	96	80	70	100 ± 6.0	12 ± 6.0	7.9 ± 0.4	0.440 ± 0.2 ^{b}	2.902 ± 1.0 ^{b}	0.201 ± 0.1 ^{b}
-S +H	100	98	70	105 ± 5.0	14 ± 4.0	8.3 ± 0.3	0.534 ± 0.2 ^{a}	3.198 ± 2.3 ^{a}	0.211 ± 0.1 ^{ab}
-S -H	98	95	68	97 ± 8.0 ^{b}	12 ± 5.0	8.0 ± 0.6	0.458 ± 0.1 ^{b}	3.013 ± 1.5 ^{ab}	0.204 ± 0.1 ^{b}

S—Na_2CO_3

H—Hoagland's nutrient solution

Mean and percentage followed by the same letter/without any letter in a column are not significantly different at 0.05 level

Also AM mycelial strands can transport physiologically significant quantities of water from the bulk soil to the root-fungus interface. The increased level of plant growth hormones in mycorrhizal association may additionally help plants under stress conditions (Ludurg-Muller *et al.*, 1997). The survivability of mycorrhiza inoculated plants on alkaline soil in present study is supported by earlier studies on mycorrhizal inoculations in different plant species (Kormanik *et al.*, 1978). Mycorrhizal inoculations are known to stimulate rooting and growth of plants (Kormanik *et al.*, 1978). Pfeiffer and Bloss (1988) succeeded in increasing the growth of *Parthenium argentatum* by inoculation with *Glomus intraradices* in a moderate and highly saline-sodic soil. Moreover AM fungi play important role in water economy of plants. Their association with plants improves the hydraulic conductivity of the roots and this improvement is one of the factors contributing towards better uptake of water by the plants. It has been pointed out that the AM fungi help the roots in better absorption of water by exploring it with wider zones of soil (Safil *et al.*, 1972). It has also been reported that mycorrhizal affected plants show better survival than non-mycorrhizal in extremely dry conditions (Allen *et al.*, 1981). Plants colonized by AM fungi have root systems, which acquire a greater percentage of photoassimilates than non-mycorrhizal ones (Clapperton and Reid, 1992). Consequently, they exhibit higher root/shoot ratios (Clappertion and Reid, 1990; 1992). By using carbohydrates for storage, for the building up of biomass, and for the conversion into metabolic energy mycorrhizal fungi create strong assimilate sinks (Dosskey *et al.*, 1990) and the degree of infection of a particular root is obviously more dependent on its carbohydrate supply. The content of soluble sugars is increased in roots or leaves of mycorrhizal plants, this could support osmotic adjustment in order to compensate for decreased soil water potential. In present study, the positive effect of AM fungi under alkaline stress situations could be due to these effects.

Also *Rhizobium* is known to survive under high salinity and alkalinity. Some *Rhizobium* species grow actively at *p*H 10 to 12 with an optimum *p*H between 6 to 9 (Horikoshi, 1991). *Rhizobia* are known to exhibit a large range of sensitivities to salinity. Moreover, rhizobia may survive in soils at salinity levels, much higher than those at which growth is restricted or eliminated in media (Singleton *et al.*, 1982). The mechanism of salt tolerance of a *Rhizobium* strain isolated from mesquite in the Sonoran Desert has been explained in terms of its ability to accumulate L-glutamate intracellularly at high concentrations. Also, *R. japoniocum* strains forming non-slimy small colonies were found to be more salt labile than those forming slimy large colonies (Upchurch and Elkan, 1977). The AM fungi and *Rhizobium* nodules in *T. foenum graecum* under alkaline stress might have helped plants under present study. The fresh and dry weight of plants decreased at high alkaline conditions (9.0 to 12.0 *p*H) in comparison to control, although no difference was observed in number of nodules between different treatments. The results are corroborated by Wilson, (1970) who reported late nodulation at high *p*H as compared to control. Also, beneficial effects of legume roots on both high and low extremes of soil *p*H have been reported (Lakshmi Kumari *et al.*, 1974). In present study, the inoculation of AM fungi and presence of *Rhizobium* in nodules of *T. foenum graecum* might have created favourable conditions for the growth of plants under alkaline stress.

The *R. communis* plants did not survive at *p*H 12.0 and also there was a significant decrease in AM spore count and AM infection percentage by increasing soil *p*H from 7.5 to 10.5 (Table 17.3). It was reported that the formation and spread of arbuscular mycorrhizas in soil is dependent on the hyphal growth of AM fungi, from germinating propagules or existing arbuscular mycorrhizas, and the subsequent infectivity of the hyphae when a receptive plant root is encountered (Juniper and Abbott, 1993). These processes and therefore, the formation of arbuscular mycorrhizas, may be affected by soil and environmental conditions, as well as by characteristics of both the fungus and the host plant. In present experiments, soil alkalinity was one such factor to affect AM infection in plant roots and AM spore count in pot soils. But depending upon characteristics of host plants, the spread of AM infectivity varied among different plant species. As in present work, the AM infection in *R. communis* was decreased more as compared to *T. foenum graecum* under alkalinity stress. The lesser mycorrhiza formation in *R. communis* may be due to effects of soil salinity on the fungus or on the plant or both. It has been reported earlier that the total length of colonization and the percentage of root length colonized by AM fungi may be reduced in saline alkaline soils (Hirrel and Gerdemann, 1980; Ojala *et al.*, 1983; Rozema *et al.* 1986). Also, dual combination of AM fungi and rhizobia in root nodules of *T. foenum graecum* increased AM infectivity better than in *R. communis*.

The Hoagland nutrient solution influenced all the parameters, i.e. AM spore count, number of nodules/plant (in *T. foenum graecum*), fresh and dry weights (both in with/without Na_2CO_3 treatments) [Table 17.2, 17.4]. The positive effect of Hoagland could possibly be due to the availability of all essential nutrients, specifically Zn for proper growth of plants and fungi, as salt affected alkaline soils are deficient

Table 17.3: Culturing of AM fungi on *R. Communis* with Na_2CO_3 (six months data, March to August, 2002)

Treatments pH	Germination (%)	Survival (%)	AM infection (%)	AM spore count/100g soil	Fresh weight (g)		Dry weight (g)	
					Shoot	Root	Shoot	Root
9.0	50[ac]	30	65	72 ± 6.0[a]	100.8 ± 2.7[a]	20.5 ± 1.2[a]	40.32 ± 2.5[a]	6.2 ± 1.0[a]
10.5	35[a]	20[a]	60	50 ± 7.0[b]	93.7 ± 2.5[b]	15.7 ± 0.9[b]	32.8 ± 2.0[b]	5.0 ± 1.2[b]
12.0	10[b]	—	—	—	—	—	—	—
7.5 (normal)	60[c]	40[b]	75	90 ± 4.0[c]	120.2 ± 3.0[c]	22.2 ± 1.0[a]	58.9 ± 2.2[c]	7.0 ± 1.0[a]

Table 17.4: Effect of Hoagland Solution on Culturing of AM Fungi in R. *Communis* with Na_2CO_3 (pH = 9.0) (six months data, March to August, 2002)

Treatments	Germination (%)	Survival (%)	AM infection (%)	AM spore count/100g soil	Fresh weight (g)		Dry weight (g)	
					Shoot	Root	Shoot	Root
+S + H	60	35	65	80 ± 6.0[a]	110.2 ± 2.3[a]	22.4 ± 0.9	52.3 ± 2.5[a]	7.2 ± 1.0[a]
+S –H	50	30[a]	60	70 ± 5.0[a]	100.5 ± 2.4[a]	20.2 ± 1.2	40.5 ± 2.8[b]	6.8 ± 1.2[b]
–S + H	65	50[b]	75	102 ± 6.0[b]	120.4 ± 2.8[b]	23.1 ± 2.0	59.0 ± 2.7[a]	7.5 ± 1.5[a]
–S –H	60	40	70	90 ± 5.0[a]	118.5 ± 2.6	22.0 ± 1.5	56.5 ± 3.2[a]	7.0 ± 1.8[a]

S—Na_2CO_3
H—Hoagland's nutrient solution

in P and Zn. AM fungi can meet the plants demand for P by solubilizing its insoluble forms but Zn is to be added.

The experiment on culturing of native AM fungi from village soil was performed. Because native AM fungi isolated from alkaline soils may be more adaptive and tolerant to alkalinity. Janardhanan *et al.* (1994) isolated a number of AM fungal spores from an alkaline *usar* land ecosystem. In their studies, native AM fungi of *usar* land ecosystem appeared to be tolerant to soil alkalinity. They recommended potential use of the native AM fungi for the revegetation and reclamation of alkaline *usar* lands. Also a consortium of native AM fungi (of *Glomus* and *Gigaspora* sp.) was used in present study, as consortium would be suitable for developing various kinds of microbial interactions and development of a healthy rhizospheric zone. A combination of fungal species, each with its own beneficial characteristics, is recommended in studies of Kernaghan *et al.* (2002) for the inoculation of seedlings growing at different *p*H levels. In the present experiment on culturing of AM fungi in village soil (*p*H 9.0), all the parameters studied (AM infection per cent, AM spore count/100g soil, fresh and dry weight, number of nodules/plant) also increased to some extent by Hoagland application (Table 17.5). Plants grown in alkaline soils often have reduced productivity because of limited mineral nutrients (Al-Karaki and Al-Omoush, 2002). The use of Hoagland nutrients solution enriched the deficient soil with useful minerals. Later on, the root based soil inoculum of AM spores, isolated from the village soil, was prepared and used for development of agroforestry models in saline-alkaline areas of Haryana.

2. Effect of Bioinoculants on Plants' Growth under Alkaline Conditions

The inoculation effect of bioinoculants (*Azotobacter, Azospirillum*) alongwith AM fungi effecting plant growth of *Morus alba* and *Populus deltoides* under alkaline conditions is depicted in Tables 17.6 to 17.9.

1. Populus Deltoides

In almost all the treatments (except control), 100 per cent sprouting of *Populus deltoides* was observed after three weeks (Table 17.6) although survival of plants was affected by induced alkalinity later on (after 5 months). It was observed that dual inoculation of AM fungi and *Azospirillum* as well as AM fungi and *Azotobacter* alone significantly enhanced growth parameters (stem height and number of leaves) and microbial count studied under alkaline village soil (Tables 17.6, 17.8) although significant differences were observed between normal and alkaline village soil.

Table 17.5: Culturing of AM Fungi on Different Host Plants on Village Soil ($pH = 9.0$)

Host plants	Germination (%)	Survival (%)			AM infection (%)		AM spore count/100g soil	
	-H	+H	-H	+H	-H	+H	-H	+H
Trigonella* foenum graecum	95[a]	98[a]	90[a]	92[a]	72	80[a]	110 ± 7.0[a]	122 ± 9.0[a]
Ricinus** communis	55[b]	62[b]	50[b]	60[b]	65	70	98 ± 6.0	105 ± 8.0[b]
Zea mays**	80[a]	75[b]	70[a]	75[b]	68	70	100 ± 8.0	108 ± 5.0[b]
Sorghum bicolor**	95[a]	100[a]	90[a]	98[a]	65	70	95 ± 5.0	102 ± 7.0[b]
Penniseteum typhoideum**	95[a]	100[a]	90[a]	95[a]	62	65[b]	90 ± 4.0[b]	97 ± 6.0[b]

* January to March, 2002 ** July to September, 2002

H—Hoagland's nutrient solution.

Mean and percentage followed by the same letter/without any letter in a column are not significantly different at 0.05 level.

Table 17.6: Effect of Different Bioinoculants on the Growth of Populus at different pH Levels 5 months Data (March-July, 2002)

Treatments		Sprouting (%)	Growth parameters			Survival (%)
			Stem height (cm)	Stem girth (cm)	No. of leaves/ plant	
At $pH=7.5$ (Micro-model soil)	Control	100	100.5 ± 2.5[a]	2.5 ± 0.5	15 ± 4.0[a]	75[a]
	AM fungi	100	120.9 ± 4.3[b]	2.6 ± 0.4	20 ± 2.0	90[b]
	Azospirillum	100	122.7 ± 3.8[b]	2.8 ± 0.3	22 ± 5.0	90[b]
	Azotobacter	100	122.4 ± 3.0[b]	2.6 ± 0.5	25 ± 2.0[b]	90[b]
	AM+Azospirillum	100	153.7 ± 2.9[c]	3.0 ± 0.4	30 ± 1.0[b]	100[b]
	AM+Azotobacter	100	142.2 ± 2.7[c]	3.0 ± 0.3	28 ± 4.0[b]	100[b]
At $pH=9.0$ (village soil)	Control	95	92.7 ± 5.3[a]	2.0 ± 0.1	10 ± 2.0[a]	60[a]
	AM fungi	100	100.8 ± 3.2[a]	2.2 ± 0.2	15 ± 2.0[ab]	65
	Azospirillum	100	102.5 ± 3.0[a]	2.4 ± 0.4	20 ± 2.0[bc]	70
	Azotobacter	100	102.8 ± 3.4[a]	2.4 ± 0.3	20 ± 3.0[bc]	70
	AM+Azospirillum	100	142.5 ± 2.7[b]	2.8 ± 0.2	25 ± 4.0[c]	80[b]
	AM+Azotobacter	100	138.7 ± 5.2[b]	2.6 ± 0.1	22 ± 4.0[c]	80[b]

2. *Morus Alba*

In *Morus alba* also, sprouting was found 100 per cent both at high (9.0) and normal pH (7.5). But survival per cent was decreased in controls in comparison to bioinoculants inoculated plants both at normal as well as high pH. Significant increase in growth parameters was also observed in the inoculated plants as compared to control (Tables 17.7, 17.9).

Table 17.7: Effect of Different Bioinoculants on the Growth of *Morus Alba* at Different *p*H Levels, 5 Months Data (March-July, 2002)

Treatments		Sprouting (%)	Growth parameters			Survival (%)
			Stem height (cm)	Stem girth (cm)	No. of leaves/ plant	
At *p*H=7.5 (Micromodel soil)	Control	100	75.2 ± 2.7^a	2.0 ± 0.7^a	20 ± 2.0^a	60^a
	AM fungi	100	110.5 ± 3.0^b	3.0 ± 0.5^b	30 ± 3.0^b	80
	Azospirillum	100	102.4 ± 2.5^b	2.5 ± 0.4	25 ± 4.0^{ab}	70
	Azotobacter	100	112.0 ± 3.2^b	3.0 ± 0.5^b	32 ± 4.0^b	80
	AM+*Azospirillum*	100	110.2 ± 3.0^b	2.7 ± 0.7	28 ± 3.0^b	90^b
	AM+*Azotobacter*	100	115.2 ± 2.8^b	3.2 ± 0.6^b	35 ± 2.0^b	100^b
At *p*H=9.0 (village soil)	Control	98	60.5 ± 3.2^a	1.7 ± 0.3^a	15 ± 3.0^a	50^a
	AM fungi	100	98.7 ± 3.0^b	2.5 ± 0.4^b	20 ± 2.0^{ab}	70
	Azospirillum	100	95.2 ± 3.0^b	2.0 ± 0.4	18 ± 4.0^a	65
	Azotobacter	100	98.0 ± 2.7^b	2.2 ± 0.2	22 ± 4.0^b	75^b
	AM+*Azospirillum*	100	102.5 ± 2.5^b	2.5 ± 0.8^b	22 ± 3.0^b	75^b
	AM+*Azotobacter*	100	105.7 ± 2.4^b	2.8 ± 0.4^b	28 ± 3.0^b	90^b

Mean and percentage followed by the same letter/without any letter in a column are not significantly different at 0.05 level (comparison was done among similar *p*H between treatments).

Table 17.8: Microbial Parameters of Rhizospheric Soil of Populus Treated with Different Bioinoculants, 5 Months Data (March-July, 2002)

Treatments		AM infection (%)	AM spore count/100g soil	Azotobacter cell count/g soil	Azospirillum cell count/g soil
At *p*H=7.5 (Micromodel soil)	Control	40^a	40 ± 5.0^a	1.0×10^2	0.7×10^2
	AM fungi	65^{bc}	92 ± 6.0^c	1.2×10^2	0.8×10^2
	Azospirillum	55^{ab}	55 ± 5.0^b	1.3×10^2	1.2×10^2
	Azotobacter	55^{ab}	50 ± 7.0^b	1.8×10^2	0.8×10^2
	AM+*Azospirillum*	70^c	98 ± 3.0^c	1.5×10^2	1.5×10^2
	AM+*Azotobacter*	70^c	100 ± 5.0^c	2.0×10^2	1.0×10^2
At *p*H=9.0 (village soil)	Control	25^a	30 ± 4.0^a	0.8×10^2	0.5×10^2
	AM fungi	60^b	70 ± 6.0^c	1.0×10^2	0.5×10^2
	Azospirillum	35^a	48 ± 5.0^b	1.0×10^2	1.0×10^2
	Azotobacter	30^a	42 ± 4.0^b	1.6×10^2	0.7×10^2
	AM+*Azospirillum*	65^b	75 ± 5.0^c	1.2×10^2	1.2×10^2
	AM+*Azotobacter*	60^b	72 ± 7.0^c	1.8×10^2	1.0×10^2

Table 17.9: Microbial Parameters of Rhizospheric Soil of *Morus Alba* Treated with Different Bioinoculants, 5 Months Data (March-July, 2002)

Treatments		AM infection (%)	AM spore count/100g soil	Azotobacter cell count/g soil	Azospirillum cell count/g soil
At *p*H=7.5 (Micromodel soil)	Control	40^a	32 ± 7.0^a	1.0×10^2	0.6×10^2
	AM fungi	70^c	92 ± 6.0^c	1.4×10^2	0.8×10^2
	Azospirillum	60^b	43 ± 8.0^b	1.0×10^2	1.0×10^2
	Azotobacter	60^b	45 ± 6.0^b	2.0×10^2	0.8×10^2
	AM+*Azospirillum*	70^c	90 ± 5.0^c	1.5×10^2	1.2×10^2
	AM+*Azotobacter*	75^c	95 ± 4.0^c	2.2×10^2	1.0×10^2
At *p*H=9.0 (village soil)	Control	25^a	24 ± 5.0^a	0.5×10^2	0.2×10^2
	AM fungi	55^{bc}	63 ± 4.0^c	0.8×10^2	0.4×10^2
	Azospirillum	50^b	38 ± 3.0^b	0.7×10^2	0.8×10^2
	Azotobacter	50^b	40 ± 6.0^b	1.8×10^2	0.6×10^2
	AM+*Azospirillum*	60^c	60 ± 6.0^c	1.2×10^2	1.0×10^2
	AM+*Azotobacter*	60^c	65 ± 5.0^c	2.0×10^2	0.8×10^2

Mean and percentage followed by the same letter/without any letter in a column are not significantly different at 0.05 level (comparison was done among similar *p*H between treatments).

In present study, overall growth of two plant species reduced by inducing alkalinity stress, was not much affected. The results have shown that combined inoculation of AM fungi and other microflora is better than the individual inoculation in improving growth and number of microbes in plants' rhizosphere both under alkalinity as well as normal soil conditions. The results are in accordance with the views of Bhanavase *et al.* (1990), Godara *et al.* (1995) and Gowda *et al.* (1995). Increased plant growth, leaf biomass and P-content in leaf were observed in mulberry plants inoculated with both AM fungi and *Azotobacter* when compared to their single inoculation treatments (Gowda *et al.*, 1995). The interaction effect of AM fungi and *Azotobacter* inoculation on micro-nutrient content of peach seedlings was reported by Godara *et al.* (1995). The synergistic role of AM and *Azotobacter* on growth and biomass production in some forestry species has also been reported by Paroha *et al.* (2000). *Azotobacter* and *Azospirillum* are known to increase the growth of the plants by means of the nitrogen fixation, production of growth regulators in rhizosphere and antagonism towards pathogenic organisms (Subba Ras, 1982). The interaction of *Bacillus polymyxa*, *Azospirillum* and AM fungi together helped in the uptake of tricalcium phosphate from soil (Ratti *et al.*, 2001). The nitrogen-fixing bacteria exploited in the present study have the capacity to grow in variable and extreme conditions. It was reported that during cell growth the bacteria could change the *p*H value of the medium to the optimum *p*H for growth (Horikoshi, 1991). It was noted that nitrogen fixing beneficial rhizosphere bacteria could interact with root exudates to produce changes in the root morphology and physiology. In dual inoculation, nitrogen fixers might have supplemented soil with nitrogen by fixing atmospheric nitrogen, which is known to increase the incidence and population of AM fungi as AM fungi has special requirement of nitrogen (Mosse, 1973). There are two factors, which have major influence on the population of nitrogen fixers in soil. These are the associative and antagonistic action of soil microflora and organic matter content of soil. Root exudates or excretions, which contain aminoacids, sugars, vitamins and organic acids together with the decaying portions of root system, serve as energy sources for multiplication of nitrogen fixers (Subba Rào, 1982). The favourable interaction of AM fungi with nitrogen fixers and use of enriched vermicompost in the present experiments might have helped in proliferation of microflora both under normal as well as alkaline conditions. The dual inoculation of these microorganisms display a synergistic effect resulting in maximum growth of plants through better N and P uptake (Habte and Manjunath, 1987) and due to secretion of phospho-enzyme and other growth promoting substance by AM, *Azotobacter*, and *Azospirillum* (Bongale and Nodiger, 1989).

The symbiotic association of nitrogen fixers and AM fungi may promote survival and plants growth under alkaline stress by certain appropriate mechanism. Various mechanisms for salt tolerance have been proposed (Levitt, 1980). The one most likely to explain genotypic differences in salt tolerant of crop plants is better regulation of salt tolerant transport preventing excessive accumulation of salt in mature leaves (Munns and Teratt, 1986). Salt resistance includes both avoidance and tolerance mechanisms (Levitt, 1980). The former may operate through active extrusion of ions by specific pumps, passive exclusion of ions due to membrane impermeability, or dilution by rapid growth associated with an increase in water content. An unavoidable consequnce of growth in a solution containing a high salt concentration is the development of osmotic stress, which is followed by a loss of turgor. Tolerance to osmotic stress may operate either through dehydration tolerance, which permits the cells to survive without growing when the turgour decreases, or by avoiding dehydration through osmoregulation, which includes an increase of solute concentration in the cells and consequently rehydration. Of all, some of the factors discussed above might have played role in enhancing all the parameters by combination of beneficial microbes under alkalinity stress.

Whichever mechanism accounts for salt tolerance, there must be an energy cost. The energy required for regulation of inorganic ions may limit the potential for plant productivity. Respiratory energy for maintenance of processes competing with growth may increase when the organism is stressed and that a greater portion of the energy is used for processes resulting in salt tolerance rather than for growth (Stavarek and Rains, 1985). The suppression of plants growth in comparison to normal soils under alkalinity in present work might be due to this factor. But the inclusion of suitable positively interacted bioinoculants maximise plant growth under alkaline stress, which may further help plants to survive in field conditions.

REFERENCES

Al-Karaki, G. N. and Al-Omoush, M. (2002). Wheat response to phosphogypsym and mycorrhizal fungi in alkaline soil. *Journal of Plant Nutrition*, 25(4): 873-883.

Allen, M. K., Smith W. K., Moore J. T. S. and Christensen, M. (1981). Comparative water relations and phostosynthesis of mycorrhizal and non-mycorrhizal *Boutelova gracilis* H. B. K. Lag ex Stend. *New Phytol.*, 88: 683-93.

Bhanavase, D. B., Mandhare, V. K., Shingte, V. V. and Patil, P. L. (1990). A study on phosphorus uptake by garlic as influenced by *Azotobacter*, VAM and P levels. *Proc. VIIIth South Reg. Conf. Microbial Inoculant*, Pune, pp. 75-76.

Bongale, U. D. and Nadiger, G. S. (1989). The role of biofertilizers in mulberry cultivation. *Indian Silk*, 28(1): 34-39.

Clapperton, M. J. and Reid, D. M. (1990). Effects of sulphurdioxide fumigation on *Phleum pratense* and vesicular-arbuscular mycorrhizal fungi. *New Phytol.*, 115: 465-69.

Clapperton, M. J. and Reid, D. M. (1992). Effects of low-concentration sulphur dioxide fumigation and vesicular-arbuscular mycorrhizas on ^{14}C-partitioning in *Phleum pratense* L., *New Phytiol.*, 120: 381-87.

Clark, R. B. and Zeto, S. K. (1996). Iron acquisition by mycorrhizal maize grown on alkaline soil. *Journal of Plant Nutrition*. 19(2): 247-264.

Dixon, R. K., Rao, M. V. and Garg, V. K. (1994). *In situ* and *in vitro* response of mycorrhizal fungi to salt stress. *Mycorrhiza News*, 5(4): 6-8.

Dosskey, M. G., Linderman, R. G. and Boersma, L. (1990). Carbon-sink stimulation of photosynthesis in Douglas—fir seedlings by some ectomycorrhizae, *New Phytol.*, 115: 269-74.

Gerdemann, J. W. and Nicolson, T. H. (1963). Spores of mycorrhizal endogone species extracted from soils by wet sieving and decanting, *Transactions of the British Mycological Society*, 46: 235-44.

Godara, R. K., Awasthi, R. P. and Kaith, N. S. (1995). Interaction effect of VA-mycorrhizae and *Azotobacter* inoculation on micro-nutrients content of peach seedlings, *Journal of Hill Research*, 9(1): 5-10.

Gowda, M. H., Das, P. K., Katiyar, R. S., Fathima, P. S., Chowdhury, P. C. and Datta, R. K. (1995). The role of synergistic effect of vericular-arbuscular mycorrhiza and *Azotobacter* in mulberry, *Mycorrhizae: Biofertilizers for the Future, Third National Conference on Mycorrhiza*, 197-201.

Habte, M. and Manjunath, A. (1987). Soil solution P status and mycorrhizal dependency in *L. leucocephala. Appl. Environ. Microbio.*, 53: 797-801.

Hirrel, M. C. and Gerdemann, J. W. (1980). Improved growth of onion and bell pepper in saline soils by two VAM fungi. *Soil Sci. Soc. America J.*, 44: 654-55.

Hoagland, D. R. and Arnon, D. I. (1950). *The Water Culture Method for Growing Plants without Soil*. California agricultural experimental station, circular no. 347. University of California, Berkeley, C. A.

Horikoshi, K. (1991). *Microorganisms in alkaline environments*. Published by Kodansha Ltd., Tokyo (Japan); VCH Verlagsgesellschaft mbH, Weinheim (Federal Republic of Germany) and VCH Publishers Inc., New York, (USA).

Janardhanan, K. K., Khaliq, A., Fauzia, Naushin and Ramasamy, K. (1994). Vesicular-arbuscular mycorrhizae in an alkaline usar land ecosystem. *Curr. Sci.*, 67: 465-69.

Jindal, V., Atwal, A., Sekhon, B. S., Singh, R. (1995). Influence of NaCl salinity on some aspects of nitrogen assimilation and mineral composition in moong (*Vigra radiata* L.) plants in the presence and absence of mycorrhizal inoculation, *J. Indian Chem. Soc.*, 42: 41-45.

Juniper, S. and Abbott, L. K. (1993). Vesicular-arbuscular mycorrhizas and soil salinity, *Mycorrhiza*, 4: 45-47.

Kernaghan, G., Hambling, B., Fung, M. and Khasa, D. (2002). *In vitro* selection of boreal ectomycorrhizal fungi for use in reclamation of saline-alkaline habitats. *Restoration Ecology*. 10(1): 43-51.

Kormanik, P. P., Bryan, W. C. and Schultz, R. C. (1978). Endomycorrhizal inoculation during transplanting improves growth of vegetatively propagated yellow poplar., *The Plant Propagator*, 23: 4-5.

Lakshmi-Kumari, M. Singh, C.S. and Subba Rao, N.S. (1974). Root hair infection and nodulation in lucerene (*Medicago sativa L.*) as influenced by salinity and alkalinity, *Plant and Soil.* 40, 261-268.

Levitt, J. (1980). *Responses of Plants to Environmental Sresses*, Academic Press, New York, 2(2): 606.

Ludwig-Muller, J.,Kaldorf, M., Sutter, E. G. and Epstein, E. (1997). Indole-3-butyric acid (IBA) is enhanced in young maize (*Zee mays* L.) roots colonized with the arbuscular mycorrhizal fungus *Glomus intraradicas*, *Plant Science*, 125: 153-62.

Mosse, B. (1973). Advances in the study of VA mycorrhiza, *Annual Review of Phytophathology,* 11: 171-96.

Mosse, B., Stribley, D. P. and Le Tacon, F. (1981). Ecology of mycorrhiza and mycorrhizal fungi. In: *Advances in Microbial Ecology* (M. Alexander, ed.), 5, Plenum Press, New York, pp. 137-209.

Munns, R. and Teraat, A. (1986). *Australin J. Plant Physiol.,*13: 143.

Murthy, R. S. (1987). *Soils in hand book of agriculture*, ICAR Pub., New Delhi, pp. 64.

Ojala, J.C., Jarrell, W. M., Menge, J. A. and Johnson, E. L. V. (1983). *Agronomy*, 75: 225.

Paroha, S., Chandra, K. K. and Tiwari, K. P. (2000). Synergistic role of VAM and *Azotobacter* inoculation on growth and biomass production in forestry species, *J. of Tropical Forestry*, 16: 13-21.

Pfeiffer, C. M. and Bloss, H. E. (1988). Growth and nutrition of guayule (*Parthenium argentatum*) in a saline soil as influenced by vesicular-arbuscular mycorrhiza and phosphorus fertilization, *New Phytologist*, 108: 315-21.

Phillips, J. M. and Hayman, D. S. (1970). Improved procedures for clearing roots and staining parasitic and vesicular-arbuscular mycorrhizal fungi for rapid asessment of infection, *Transactions of the British Mycological Society*, 55: 158-61.

Ratti, N., Kumar, S., Verma, H. N. and Gautam, S. P. (2001). Improvement in bioavailability of tricalcium phosphate to Cymbopogon martini var. motia by rhizobacteria, AMF and Azospirillum inoculation. *Microbiological Research.* 156(2): 145-149.

Rozema, J., Arp, W., Van, D. J., Van, E. M., Broekman, R. and Punte, H. (1986). Occurrence and ecological significance of vesicular-arbuscular mycorrhiza in the salt marsh environment. *Acta Botanica Neerlendica*, 35: 457-67.

Safir, G. R., Boyer, J. S. and Gerdemann, J. W. (1972). Nutrient status and mycorrhizal enhancement of water transport in soybean. *Plant Physiol.* 49: 700-3.

Singleton, P.W.E.L., Swaify, S.A. and Bohlool, B.B. (1982). Effect of salinity on *Rhizobium* growth and survival, *Applied and Environmental Microbiology*. 44, 884-890.

Stavarek, S. J. and Rains, D. W. (1981). Ion uptake of salt selected plant cells. *Proceedings, 57th Annual Meeting of the American Society of Plant Physiologists*.

Stavarek, S. J. and Rains, D. W. (1985). In: *Cellular and Molecular Biology of Plant Stress* (J. L. Key and T. Kosuge, eds.) Alan R. Liss, New York, p. 129.

Subba Rao, N. S. (1982). *Bioferilizers in Agriculture*, Oxford and IBH Publishing Co. Pvt. Ltd. New Delhi, India.

Upchurch, R.G. and Elkan, G.H. (1977). Comparison of colony morphology, salt tolerance, and effectiveness in *Rhizobium Japonicum*, *Canadian Joiurnal of Microbiology*, 23, 1118-1122.

Wilson, J.R. (1970). Response to salinity in Glycine VI. Some effects of a range of short-term salt stresses on the growth, nodulation, and notrogen fixation of *Glycine Weightii* (formerly *javanica*), *Australian Journal of Agricultural Research*, 21, 571-582.

Yadav, V., Madan, L. and Jaiswal, V. S. (1990). Micropropagation of *Morus Nigra* L. from shoot tip and nodal explants of mature trees. *Scientia Hortic*. 44: 61-67.

IV
Water Resources Management

18

Restoration of Tanks for Water and Ecological Security: An Ecosystem Approach

C.B.S.R. SHARMA

INTRODUCTION: HISTORY AND THE CURRENT STATUS OF WATER BODIES

Water, that comes through rainfall is ultimately garnered in wetlands and tanks thereby influencing human settlements since the time immemorial. In the face of current water scarcities, revival of these traditional storage systems becomes necessary for a fuller utilisation of rainfall. This revival can be brought about by conserving and developing the 2,00,000 tanks spread across Tamil Nadu, Andhra Pradesh, Karnataka, Maharashtra and parts of Rajasthan, Bihar, and Uttar Pradesh. Unfortunately, there is neither a comprehensive data base on them nor an in-depth study on anyone tank until very recently (Sharma, 1996, 1997a, 1997b). However it does not mean that the tanks are altogether ignored. The Second Citizens' Report on the state of India's Environment tells us that only about 10 per cent of the annual precipitation is utilised for want of proper storage systems underscoring the value of ponds and tanks in water management.

Periodical Concerns of the Governmental Agencies

The Famine Inquiry Commission (1944) recommended construction of tanks and their proper maintenance through appropriate legal inputs.

The Grow More Food Campaign of 1943 which was subsequently integrated into the First Five Year Plan recommended greater attention to minor irrigation. The Second Irrigation Commission also surveyed the status of tanks in India. Yet the number of tanks and the acreage irrigated by them consistently dwindled all over the country (Rao, 1979; CSE, 1984). Bhumbla (1982) recommended specific practices of management of these resources in about 300 districts in India for augmenting groundwater and helping subsistence farmers. But these necessitate comprehensive water collection and distribution, soil conservation and habitat vegetating. Yet planners and policy makers seem to be unconcerned in spite of telling figures, periodical droughts, regional shortages and ominous portents.

Global Ecological Perspective

The International Union for Conservation of Nature (IUCN), World Wide Fund for Nature (WWF) and Ramsar Convention (1971-89) recognised the importance of wetlands—reclamation and management of water bodies (Maltby & Turner, 1983; Derek & Colin, 1989 and Scott & Poole, 1989). Therefore, the Government of India formed a National Commission on Wetlands (GOI, 1989) to be aided by State Level Steering Committees based on the feedings from the Area Development Authorities. The various management aspects, over a dozen, were to be implemented and managed through four Acts of parliament—The Wetland Conservation Act (1972). The Water Act (1974), The Forest Conservation Act (1980) and The Environment Protection Act (1986).

While IUCN (Dugan, 1990) has detailed criteria for wetland identification and management, the Indian Government detailed the needs of research on a score of issues and has even settled for a management plan. This helped raise data on some important wetlands, but none on the vast mass of small waterbodies which include tanks.

The WWF had listed and classified about 100 wetlands in 1989, which included a few irrigation tanks and estimated the wetland area to be about 3.0 mha which is nearly 100 per cent more than the previous estimations. The first among the nine types of wetlands in India, interestingly, happens to be tanks, reservoirs and other similar water bodies. These differed in size and shape depending upon the topography, and are mostly under local organisations and powers such as landlords and zamindars.

In all about 400 species of fish alone, flourish here. Further, they are winter resorts for birds; they harbour reptiles, mammals and other lower fauna; provide food and fodder; help in flood control and production of organic material. They are natural filters and regulate water regime. These are supposed to be the most productive systems in both the submerged and other macrophytes (MOEF, 1992).

Status of Indian Tanks

The oldest man-made tank in India is supposed to be 20 centuries old, near Allahabad, not considering the tanks of Indus Civilisation. In Bellary and Telengana some tanks date back to 12/13th centuries. They are situated in different agroclimatic zones—coastal districts of Tamil Nadu and Andhra Pradesh, South Central Karnataka, Telengana, Eastern Vidarbha, North-eastern Uttar Pradesh and Eastern Rajasthan. They were managed by a variety of traditional practices. But most of them are now dysfunctional although they are a low investment and locally manageable irrigation systems (Chaturvedi, 1968).

There is no comprehensive and reliable database on the tanks, for monitoring and making appropriate and developmental inputs. On the other hand, the land value of the tank is greater wherever there is a concentration of habitations, and is, therefore, invariably encroached upon. Gradually the water bodies have been filled up and built upon. Foreshores have been encroached. Tubewells have increased multifold and are more attractive, as the common property of tanks became systematically difficult to maintain. Droughts, siltation, absence of zamindari system and non-functional Panchayats have added to the disuse of tanks even though they are economically profitable and ecologically resourceful for drainage, storage, charging wells and reducing flood havocs (CSE, 1984-85, Sharma, 1997).

In some districts of Andhra Pradesh, tanks disappeared along with Zamindaris. In Gujarat only 20 per cent of a total of 1142 tanks surveyed were in good shape in 1964-65. In Madhya Pradesh 26.4 per cent of irrigated area was accounted for by tanks in 1952 dwindling to a half by 1968-69. In Vidarbha the acreage under tank irrigation has fallen from 1,57,000 ha in 1954 to 97,000 in 1960. In Karnataka while the irrigatable area increased from 5,05,000 ha in 1905 to 12,34,000 ha in 1968-69, that of tanks increased only by 91,000 hectares from 2,60,000. The erstwhile Madras Presidency had about 53,000 tanks accounting for 50 per cent of irrigation source, which however, dwindled to a mere 10 per cent even while the cropped area increased eight fold. In Telengana the acreage under tanks increased from 4,000 ha in 1895 to 3,64,000 ha in 1940s. The national coverage however decreased from 4.8 mha in 1958-59 to 3.9 mha in 1978-79. Tanks in India until recently account for 11.7 per cent of irrigated area, i.e. 3.3 mha (Rao, 1979; Nadkarni, 1979; von Oppen & Suba Rao, 1980; Guhan, 1984; Mukundan, 1988).

Attendant Plagues

It has been known that all sectors/resources of a wetland/tank are interconnected among themselves and with the rural society as a whole. The tank land and its water are the most important of these common property resources. With the decline in tank rights, duties and managements for reasons of planning and politics, water and land were gradually and systematically owned up.

Water Markets, Tyranny of the Borewells

The water residence in tanks is reduced due to tube well extraction, excessive opportunity usage, etc. The tube wells around tanks extract the ground water which is replenished by the surface water in the tanks due to soil dynamics. This constitutes an ownership of the common surface water as private groundwater and sold to neighbouring farmers for a profit. And the electrical charges to privately extract this common water is subsidized—a case of Governments abetting destruction of common properties ! Consequently water-less geographical spaces are created which are readily encroached by the locally powerful—mostly the owners of the borewells. Inequalities are generated or stabilised. Borewell policy, therefore, should be based on the dynamics of surface and underground water—the location and extent of aquifers. Otherwise one will recharge and others will extract, consume and sell.

Encroachments/Land Grabbing

With the absence of managerial systems replacing the traditional ones, and the spaces created by borewell extraction, tank lands are encroached by the neighbouring land owners, local landlords and political forces. Sometimes the government itself "encroaches" on the land in the name of social forestry [which should actually be called wetland forestry]. Consequently many tanks are mere apologies, as they appear no more than temporary ponds. This anarchy is beneficial to a few and detrimental to most.

However, there is an interesting variation to this common trend. In Kaliveli, of South Arcot district, the offshore is encroached by a large number of villagers, apparently from the lower economic order. And the government has gradually been legitimizing these by awarding 'pattas'. This is expected to be inconsequential due to the vastness of the wetland. But for how long this socialism through shore-land reforms at the cost of water ecology survives? Perhaps until the Aquaculture farms arrive. And they arrived now on the quiet (Sharma, 1997).

THE CURRENT FOCUS: PARADIGM SHIFTS

Protection and regeneration of water bodies has now emerged as an urgent imperative. This can be best done by their beneficiaries and users by helping them develop stakes in these systems. This is possible by two approaches:

1. *Ecosystem approach to irrigation tanks*: Villages and every-thing around the tank are to be considered as parts of the tank system and not *vice versa*. Therefore, everything needs conservation and development in a holistic fashion and not sectorially.
2. *Treating all water bodies in one genre, one umbrella, one taxonomy and extending the system approach to them*: Wherever water can be impounded from loss by whatever means or policies it should be done, thus extending the approach to all water bodies.

Irrigation Tanks are Ecosystems

The wetland tank ecosystems are constituted by soil, water, plants and animals, i.e. habitats and biodiversity in and around the waterbody. These attract human settlements and with them agriculture and aquaculture. The water dynamics encompass catchments and command, inlet and outlet channels, bed and shorelines, embankments and slopes, wells and borewells, agriculture, aquaculture, animal husbandry, social forestry villages and inhabitants as well as impediments and adjuncts to these. These systems have some features of dammed reservoirs, natural lakes, shallow wetlands and riverine shorelines. Consequently, subsystems and micro-habitats vary from one another among tanks(Yoganand *et al.*, 1993).

But in all the tank areas, the waterbody functions as a 'cornerstone' component of the region, drawing the surrounding villages for their water, soil, plant and animal needs. In other words, the villages depending on this 'epicenter' appear as a part of the tank ecosystems and not *vice versa*. Thus, if the tank is not there, many villages will not exist. This is also borne out from history. Human settlements are known to have disappeared along the ancient Mangamma Salai in Southern Tamil Nadu because the tanks seem to have run out of water continuously for some years (Venkatraman and Sharma, 1983). Modern science would have averted such mishaps.

There are many wells around the tank—conventional and bored, reducing the direct dependence of villages on the tanks. But the wells are dependent on the tanks for their lateral and/or geological recharge. In this connection it is necessary to study the dynamics of these recharges for proper location of wells, the groundwater potential, its potability, level of water table and potential for recharge from tank waters. The dynamics of tank-well interactions is a fascinating study of geoecology ignored so far. Since the villages are key constituents of the above web both tank conservation and village development have to go hand in hand.

All Water Bodies Under One Taxonomy

Extension of the above approach to wherever and whenever water can be impounded by whatever means or policies and methods is a natural corollary of the ecosystem approach. Thus coastal tanks and ponds, urban tanks, temporary ponds, perennial private ponds of Orissa, Bengal and Assam, Sacred ponds of Temples and Gurdwaras and village ponds of deltaic systems and others are to be covered by this system approach. While the number of irrigation tanks are expected to reduce from 2,00,000, the other kinds of water bodies are expected to add as many if not more.

As per the paradigm shift mentioned above not only the number but also the variety of water bodies dramatically enhances. This calls for a taxonomic approach—classifying and characterising them based on some criteria for which an attempt is made herewith.

1. Irrigation Tanks: Types.—System and Non-System

These are the typical tanks whose characteristics and history have so far been dealt with. The system type have engineered structures for everything while the non-system tanks entirely dependent on the local rainfall and catchments.

2. Coastal Tanks: Types—Occurring Very Near to the Coast

Some are same as above. Some have Lotus populations. The others are shallow Subsurface Oozers—These are small ponds forming with the removal of a few feet of sandy soil, during monsoon and post-monsoon periods, as if water is oozing (Ilangovan *et al.*, 2002).

3. Temple Tanks: Types—Internal and External

The places of worship of many religions have tanks as parts of them—Hindu Temples, Sikh Gurudwaras etc. These may be situated within the temple premises or outside it, which make a difference in management.

4. Perennial Monsoonic Ponds: Types—Home Stead and Farm Stead

In many deltaic regions and in the states of Orissa, Bengal, Assam etc. there are private ponds nearer or around the home steads or in their agricultural fields. These serve the relevant local needs.

5. Urban Tanks: Types—System and Non-System

These are basically legacies or vestiges of the past. Due to population growth and urbanisation they have transformed a great deal from their original structures and functions. Some of them are entirely dependent on the local catchment while a few still can receive water from supply systems. Consequently, the characters, problems and values of urban tanks are different from the rural ones. Conservation or abandonment are the choices here while in the rural systems it is either restoration or conservation.

Wetlands seem to have become the easiest and popularly justifiable victims of private property and speculative profiteering. In this connection one may recall instances of "converting water into soil" in many small towns by dumping solid wastes for "reclaiming". This owned up land is used for raising concrete jungles. Consequently drainage of rain water was reduced and charging the underground aquifer hampered. What is the gain? What is the loss? And for whom?

VALUING THE FACILITIES FOR FURTHER USE

It is well known that tanks have been long-standing systems. Irrigation tanks alone are about 2,00,000. Inclusion of other types enhance this to 5,00,000. But no one knows howmany still exist on ground, in what condition and various inputs they require to function, if at all. Accounting all resources harboured here and valuing them on relative terms may give us an ecological and economic perspective for development options.

Therefore, there is a need to assess and decide on their fate. This can be approached in three ways: (i) Census and Diagnosis, (ii) Valuation and Decision, and (iii) Rapid Evaluation

(a) Census and Diagnosis

Census involves the most basic data: Does the tank exist? Is it used by the people? How are the features of the tank compared with past records?

Diagnosis deals with details: Is the tank encroached? Totally, mostly, partly, slightly, inlet side, outlet side, periphery, bed and seasonally or permanently? Is it used for irrigation or for something else? How are the banks, supply channels, sluices and ayacut functioning? What is the geography of the ecosystem and area of the tank and its water holding capacity? What is the nature of soil, water, flora and fauna?

(b) Valuation and Decision

Valuation deals with the utility of the tank to society and mainly deals with the structure, the user value and attitude of the people which are ascertained by the following questions: What are the inferences from the census and the diagnosis? Do they suggest alternate uses or continuation of the tradition? What are the current uses of the tank? Is there social or wetland forestry? Is the tank valuable or useless and why? How do local people perceive on the utility of the tank?

Decision on the fate of the tank depends on the answers to the above questions. The answers are to be given a value rating, pooled and arranged on a scale of 1-10. Higher rating reveals good health and lower rating sickness. Based on this decision is to be taken whether to conserve, restore, or abandon the facility. The methods of valuing and scaling are at conceptual stage for now. The above value system will gradually evolve revealing indicators, correlations etc, and become speedier through practice.

(c) Rapid Evaluation System

Yet the problem remains as to the methods to be followed for valuing such large numbers very rapidly, lest the efforts remain a mere academic exercise. Large numbers involve setting priorities which necessitate a rapid evaluation of the state of a tank which can be done only by two methods: (i) Direct Survey as detailed above by an army of NGOs wherever tanks exist and (ii) Developing an Expert System.

Developing an expert system for diagnosing all the irrigation tanks is a project with enormous significance in the present context. Further, this exercise will result in the compilation of all available data and knowledge base on the tanks for further research. Nowhere else one can witness an interplay of ecology, economics, sociology and politics. Some attempts are already made, but not yet invoked in the field (Oswald, 1991).

ACCOUNTING RESOURCES AND STAKE DEVELOPMENT

(a) Habitats: Land, Soil and Water

The quality and resourcefulness of the land mass around the tank depends upon topography, physical, chemical and microbiological properties and land use which also influence the quality and quantity of water. These need to be assessed along with the rainfall, catchments, run-off, storage and seepage, groundwater recharging to enable a better stock taking of water resources. Most land must be used for cultural purposes and a minimum for constructions. Bird and animal excreta should be used as fertilizers. It is a pity that the coastal rainfall is not prevented from flowing into the sea. The systems are amenable to both traditional and modern science, a blend that is warranted for the restoration of lands around tanks.

(b) Biodiversity: Plants and Animals

Two types of plant resources occur around tanks—cultivated and natural. Traditional cultivars may possess special characteristics—such as resistance to pests and drought—a very valuable germplasm for further use. Natural flora may contain plants useful as fodder, medicines, hedges, soil binders, fuel, food, fibre and fertilizer. Replacing them with firewood monocultures is not wisdom.

There are many wild animals in the wilderness as well as domesticated animals in the homesteads. These consumers of water and plants are resources to man and land. Fish are a natural source consumed locally and/or sold. Some wild animals are hunted for food or trade. Animal excretions especially of birds, guano, add great fertility to soil (Amrut *et al.*, 1998, Manimekhalan *et al.*, 2002).

(c) Resource Inventory

Surveys of land, water, plants, animals and human practices around the tanks result in checklists of these which enrich the data bases of the national and regional agencies. But such inventories are available for less than 10 tanks so far (Sharma, 1997) out of 2,00,000 tanks of India associated with thousands of villages and millions of people. This neglect appears to be the chief reason for lack of rural development.

(d) Local Perceptions and Initiatives

There are villages around the waterbodies with varying populations and characteristics. Interactions of these people with the local resources depend on their perceptions of them and the initiative they are willing to take psychologically. Consequently, there are wide variations in the tank conservation and rural development. Most lands were encroached and reducing them to vestiges of their past. In others agriculture, aquaculture, social and wetland forestry besides direct use of tank soil, water, plants and animals continue. Some of these are considered here.

(i) *Agriculture and Ecoagriculture*

In rainy season, rice is planted in most tank-fed areas but using channeled or ground water. Fertilizers and pesticides follow. Sometimes sugarcane is cultivated. This scenario is unsuitable to the semi-arid areas where most irrigation tanks exist. Here the emphasis should be on water economy, soil conservation and equitable water sharing. Water should be retained for long in the tanks and wells but not privatised through tubewells. In other words, there should be farming of crops needing minimum water per unit area which additionally bring more land area under cultivation.

In such cases it will be easier to develop ecologically sound agriculture on a long term basis. Strategies like mulching, composting, growth of biofertilizers, vermiculture, pest control through biological means, usage of animal excreta especially guano, can all be initiated and maintained as needed for sustainability.

(ii) *Aquaculture: Capture and Culture*

Capture fisheries have been found to gradually dwindle which is bad for the local poor people. The waterbody should be fully utilised for natural growth of useful fauna with inputs of animal seed and plankton. Economics of these aspects need investigation for its revival as a poverty alleviation measure. The existing scenario is very negative. Tubewells take away the tank water needed for fisheries. Avarice of the privates pirate common properties and their ecological contributions to the community.

(iii) *Sericulture, Apiculture, Floriculture, Pomoculture and Pastures*

Sericulture has two components—the silk moth and the mulberry. Tanks can support mulberry crop on its bank slopes and elsewhere fertilized by the soil from tank bed or shoreline. When these support a legume forestry or function as a bird sanctuary the soil utility is greater. The Chinese experiences in this regard are very relevant here. On the other hand apiculture can be practiced only if biodiversity is greater and vegetation is abundant. In semi-arid regions of India where most irrigation tanks occur, vegetation is sparse. When tanks are restored, flora and fauna also enhance facilitating honey production by the apiaries.

Flowers and vegetables can also be raised seasonally and fruit crops on perennial basis. Suitability, however, differs from tank to tank primarily on edaphic factors. Tank shores, banks and its slopes can support growth of grass selected for grazing by cattle.

(iv) *Tree Cultures: Wetland Forestry*

As for the tree cultures, it should be differentiated into plantations within the tank and outside it. Those within are inundated by water for various durations and serve as animal habitats and termed as wetland forestry. Those outside are plantations on wastelands receiving only natural rainfall. The former must be regulated because they yield greater biomass, enrich soil fertility and provide habitat for animals; but cause gradual reduction in tank area which however can be taken care of through engineering works (Sharma, 2003).

Silvicultural practices using leguminous plants in the tank shores improves soil quality and diversity and richness of animal populations. But the production, marketing and managements need detailed inquiries as this is a rare case of doing the best in the worst of times! The economics seem to have been totally ignored wrongly assuming that silviculture in the tank bed is equal to the social forestry outside it.

1. Bird Sanctuary

All the tanks are a sort of bird sanctuaries already to different degrees and durations. The ecology of birds have influence on every other life through food chains, reproductive necessities, nutritional ethologies etc. some of which are of economic significance to the ecosystem. The excreta of birds are of variable quantity and contribute to fertility. They bring in plankton from other water bodies. And some of these are predators on the insect pests of crops.

2. The Guano of Birds

Because of trees more birds are attracted, though seasonally. This encourages poaching for local consumption. The excreta droppings gradually change the soil nature along with fallen leaves. It becomes more fertile and biologically active supporting a little herbaceous and ground cover. In a few years the soil becomes all the more fertile. This soil is often extracted, but sparingly, for use in the local agriculture.

3. Water Quality

The quality of the water changes. With natural addition of nitrates and phosphates water becomes potable. In agriculture chemical fertilizers become avoidable. If such supplements are added there is a reduction of seed yield and increase of vegetative growth. In nearby places the ground water becomes tastier, as claimed by some local people in some places.

4. Stake Development

This experience made the local people wiser. The tank becomes a precious commodity to be protected. Private prosperity from this commons is discouraged—pumping water out, sinking borewells around the tank for geological withdrawal of surface water, etc. Utility-oriented desiltation and understanding of ecological phenomenon seem to have begun. Proper trees at proper places lead to a series of changes in due course of time—Birds and their Guano—fertility of soil—fertility of water—potability of water—substitution of chemical fertilizers—protection of bunds—enhanced water holding capacity etc.

All tanks will be known for their characteristics and values. Their further improvements depend on these. Thus each tank not only takes the local water shed but also serves the local villages in several ways—This tank dependent sectorial development renders the tank a stake for the water and soil security in the region.

5. Conservation of Common Property—Prevention of Private Exploitations

Needless to emphasize that local rainfall can be best harnessed by this indirect way of routing it to tank-trees than allowing to run off to the sea as in the coastal regions. Tanks thus become a valued interest with people as it was centuries ago. Then we can say that ecological wisdom has come to life once again.

Where the government, academicians and NGOs do not succeed enough to motivate local people to take care of their common tank facility, the legume trees and the birds on them have facilitated. In fact the local people seem to develop a stake in this system gradually.

6. Relevance to Several States Especially Coastal Regions

There could be lessons from these to the tanks in the country. Many of these will be amenable for this wetland forestry practice. There are many other types of wetlands all over the country which can be engineered for the above practice in specific locations, especially at the shorelines. This could be a very good augury for water bodies of the coastal and deltaic regions.

PERSPECTIVES: THE FUTURES

(a) Actions on the Ground and Policy Inputs

The tank may be useful in many ways: potable water for humans, potable water for animals, irrigation of crops, irrigation of horticulture gardens, habitat for aquaculture, conversion as sewage oxidation pond, washing kitchen ware, clothes and cattle wetland or social forestry, charging ground water table, bird sanctuary, collection of birds excreta, sericulture, apiculture, nursery, hatchery, tourist spots, etc.

The villagers know the importance of tanks in their economy but not their ecology—an impediment to their effective participation in sustainable utilisation of the tank resources. Sound actions on the ground are as follows: Strengthening of embankments, repairing regulators, sluices, supply and feeder channels, desiltation of the waterbody, preventing poaching, giving fishing and grazing rights and removal of encroachments. Regulating the land use in the surrounding areas of tanks so that all rain water drains into the tank. Establishing initiatives in favour of organic farming, rotation cropping, medicinal plant gardens, wetland forestry, aquaculture, etc. and educating the local people on all the aspects of the tank system (Baliga, 1960; von Oppen & Subba Rao, 1980; Palanisamy & Easter, 1984; Janakarajan, 1991).

Policies should be based on sound ecological wisdom. Agricultural policies regulate land use through revenue subsidies, extension and irrigation. Water intensive crops and bore wells in tank vicinity should be restricted through fiscal policies. Drought resistant plants can be introduced in agriculture and social forestry.

(b) Temples and Tanks: The New Focus

1. Temples are the sacred places of worship to Hindus. They are everywhere and in fact countless. So countless that even the 30,000 temples destroyed during the Muslim Rule and deities transferred as foot steps of Mosques in other countries, constituted only a miniscule. Though their lands are swallowed by the land sharks, tenants don't pay the revenue, even if the premises are encroached, yet temples are still venerated and some semblance of worship continues. Though the traditional priesthood is vanishing especially in South India, leading to their closure, yet some ingenious methods are resorted to see them alive (Sharma, 2003).
2. Despite the pluralism within Hinduism—of Gods, temples, properties, confiscations and appropriations, there is trust in the institution of temples and their sanctity. This can be exploited for the common worship of the new God—Ecology—as it is synonymous with the safety of nature. A

psyche of respect to the elements—water, plants and animals here—through the agency of God—can be inculcated, however, agnostic or atheistic one may be.

3. In case of the existing temples tanks can be desilted and developed. If the tank is outside the premises of the temple, maintain fast growing and slow growing plants and grass—temple forestry—in earlier parlance—sacred groves. Rain water can be channeled into this. In these sacred groves and sacred tanks the devotees can maintain locally important plants and animals, for the sake of God and his children's common welfare.
4. In case of the existing tanks, about 2,00,000, construct a temple on the banks of the tank if necessary. In the process the tank becomes temple's property and both together custodians of the region's welfare to some extent. Desilt the tank and build its embankment, grow grass, herbs, shrubs and trees of slow and fast growth based on the local knowledge. A sacred grove and a sacred tank will thus emerge in due course. Both serve as additions to local people's natural resource. Thus all temples get tanks and all tanks temples, both to be maintained by the local people in their own interests.
5. In fact places of worship of all religions must develop a sanctum for water and trees in their premises or neighbourhood. I saw a mosque doing this on the southern east coast of Tamil Nadu. Sand dunes of east coast are being stablised and improved for the habitat and people by the church. Churches and Christianity have a programme for nature conservation. Why not all others emulate?
6. There could be many fringe benefits—local recreation—parks, bird sanctuary, increased fertility, unexpected income from fisheries, tourist attraction, etc. in due course of time. These, therefore, can function as biodiversity refugia. It is unfortunate that the village tank of India didn't find a place either in the wetland convention of Ramsar or in the development alternatives in the literature.
7. Many of these suggestions have been a part of our traditions. Therefore, the present proposal is a mere update of the local ethos and wisdom. Maybe this can be a part of Janmabhumi Programme of Andhra Pradesh. It is for the local authorities to decide on the rules and regulations and modalities

governing the above scheme. If necessary a law can be passed to implement this plan—this model for resurrection of the Mother Earth and development of the local peoples' security consciousness.

8. In other words, the lands and water bodies belonging to the places of worship of all religions in India need restoration from an ecological point of view. Since these are the times of remembering Gods and religions, government should use their revivalist psyches for using their resourcefulness for people.

By this method the number of new and better tanks due to the existing temples, and the number of conserved tanks due to new temples will rise many fold, a good augury for water security in some ecosystems of India.

(c) Empowerment of the Beneficiary: Panchayati Raj and the NGOs

Tank irrigation requires a strong local authority to generate collective action for tank building, to efficiently distribute the water by optimally allocating it among beneficiaries and over seasons and to properly maintain the irrigation system. Tank irrigation thrived till a few decades ago because water distribution was carried out by local bodies of farmers in the ayacut or the zamindars. Now farmers do not have any say. Tail-end farmers had to shift to groundwater irrigation. This can be reversed only by reviving the traditional water management bodies.

The water dynamics encompass the entire catchment, supply channels, command area, the wells, impediments and adjuncts to these. The villages are intimately connected with this web as a part of it. Therefore, both conservation and development have to go hand in hand by nucleating each tank along with the dependent villages into a unit—An Ecopolitical unit. Empowering this unit with the responsibility to regenerate the waterbodies, catalysed by political power. However, this cannot happen unless appropriate NGOs are invoked in these efforts.

Non-Governmental Agencies (NGOs) should help to form co-operatives to centralise all operations involving some of these: vermiculture, sericulture, apiculture, floriculture, plant and animal refugia, gardens and above all water sharing. On one hand, people can take up developmental activities yielding produce and income. To sustain this for long, on the other hand, conservation measures of water, soil, plants and animals become prerequisites.

(d) Tank Rehabilitation Programmes of the EU and the River Links

Restoration of Tanks must involve all approaches—Engineering, Societal and Ecological. First and the second are employed and the third is neglected in the tank rehabilitation programmes in Pondicherry Union Territory and Tamil Nadu financed by the European Union. These EU projects also ignore a relevant and useful report of the UNDP-DOEF-MIDS sponsored research project. Conservation and restoration on long range basis are therefore not ensured. Thus there is a danger of these faulty programmes becoming models for other states (GOP, 1989; Sachuda and Sharma, 2002).

Concerning the river linking pilot works have been taken up, e.g. Pennar-Palar-Kaveri], associating these with the existing tanks, or making new ones is a natural corollary to the alignment of the link canals. Ignoring this now will entail greater costs subsequently.

(e) Updating the Records, Creating Records

The data on the tanks given by the Revenue, Irrigation and Agriculture Departments differ from the ground realities for a variety of reasons—natural degradation (siltation, aridity) man-made degradation (encroachments) and accent on alternatives (wells). Further, how much is the rainfall? How much is retained? What is the area of the tank originally and now? How much is encroached and legalized? What is the water spread area and volume of water it can hold now? Therefore, examining the characteristics of tanks and a revision of governmental statistics on the tanks is very urgently needed.

The new tanks suggested around places of worship and details about the other types of tanks have to be integrated with those of irrigation tanks. A formidable task indeed is awaiting us.

(f) National Tank Authority

The subtle interrelationships among agriculture, social forestry, land and water dynamics, biodiversity, the socio-economic mileu and above all the political undertones running through the entire system of waterbodies have to be realised if they are to be resuscitated. Since these waterbodies are a major group of natural, irrigation and drainage sources, there is a need to change our approach towards their management radically, i.e. from an Engineering outlook to an Ecosystem Approach. This can be done by the establishment of a National Tank Authority which should be mandated to take care of all types of tanks in India both for conservation as well as development on an ecologically sound manner (Sharma, 1997a).

REFERENCES

Amruth, M., Yoganand, T. R. K. and Sharma, C. B. S. R. (1998). "Preliminary Studies on the ecology and biodiversity of Ousteri and Kaliveli, two wetlands near Pondicherry" (in) *Current Researches in Plant Sciences* (ed. Sarma T. A. *et al.*): 353-358.

Baliga, B. S. (1960). "Irrigation tanks and their problems". In: *Studies in Madras administration*, 2: 199-208. Government of Madras, India.

Bhumbla, D. R. (1982). *Small Reservoirs: A Programme for Improving Rainfed Agriculture*, Society for the Promotion of Wasteland Development.

Conservation of Wetlands in India, (1989). Ministry of Environment and Forests, Government of India, New Delhi.

Chaturvedi, B. V. (1968). The origin and development of Tank irrigation in Peninsular India, *Deccan Geographer,* Vol. VI.

CSE, (1984-85). *The State of India's Environment*. The Second Citizen's Report, New Delhi, p. 393.

Derek, A.S and Colin, M. P. (1989). A status overview of Asian Wetlands, *Asian Wetlands Bureau*, Malaysia.

Dugan, P. (1990). *Wetland Conservation, A review of Current issues and Required Action*, IUCN.

Government of India, (1989). *Conservation of Wetlands in India*. Ministry of Environment and Forests, p. 67.

Government of Pondicherry, (1989). *Modernisation of Tank Irrigation Systems in Pondicherry Region of Union Territory of Pondicherry,* Public Works Department, Vol. 1: p.125.

Guhan, S. (1984). *Irrigation in Tamil Nadu*: A Survey, Madras Institute of Development Studies, Madras, pp. 75.

Ilangovan, K., Debashish Patnaik, Kavita, D, Sudeep, P. Smita Krishnan, Mary, M.S. Anbalagi, S. Muthuandavan, L. and Sharma, C. B. S. R., (2002). "Reconnaissance on the East Coast of Tamil Nadu—Survey on the status of wetlands and potable water for evolving conservation and development strategies". [in] Symposium on Conservation Restoration and Management of Aquatic Systems, Indian Institute of Sciences, December 9-13, 2002, p. 108.

Janakarajan, S. (1991). *In search of Tanks: Some Hidden Facts*, Madras Institute of Development Studies, Madras, p. 38.

Maltby, E. and Turner, R. E. (1983). 'Wetlands of the world', *Geographical Magazine*, 55: 12-17.

Manimakalan, A., T. R. K. Yoganand, J. Rajan, M. Amruth, K. Ilangovan and C. B. S. R. Sharma (2002). "Biodiversity and Ecology of two endangered wetlands of Tamilnadu Coast—Kaluveli and Ousteri. [in] Symposium on Conservation Restoration and Management of Aquatic Systems, Indian Institute of Sciences, December 9-13, 2002, p. 60.

Ministry of Environment and Forestry, (1992). *Conservation of Wetlands in India*, GOI, New Delhi, p. 48.

Mukundan, T. M. (1988). The Ery Systems of South India. *PPST Bulletin No. 16*, Madras.

Nadakarni, M. V. *et al.* (1979). *Impact of irrigation: Canal, Well and Tank irrigation in Karnataka*, Himalaya Publishing House, Bombay.

Oswald, O. (1991). *An expert system for the diagnosis of tank irrigated systems*. Institut Francais De Pondicherry, p. 157.

Palanisami, K. and Easter, K. W. (1984). "Irrigation tanks of south India: Management strategies and investment alternatives", *Ind. Jour. Agri. Econ.* 39: 214-223.

Rao, K. L. (1979). *India's Water Wealth*. Orient Longman, pp. 267.

Sachudha, V. and C. B. S. R. Sharma (2002). "Applying Ecology in restoring tanks under Tank Rehabilitation Project". [in] Symposium on Conservation Restoration and Management of Aquatic Systems, Indian Institute of Sciences, December 9-13, 2002, p. 141.

Scott, D. A. and Poole, C. M. (1989). A Status Overview of Asian Wetlands, No.53. Asian Wet Bureau, Kualalumpur.

Sharma, C. B. S. R. (1996). "Ecological Revival of Tanks and Ponds". *The Hindu*, 26, December.

Sharma, C. B. S. R. (1997a). *Ecography of Irrigation Tanks. A pilot study of six water bodies in the semi-arid Tamil Nadu and Pondicherry.* Madras Institute of Development Studies. p. 129.

Sharma, C. B. S. R. (1997b). Resurrecting a Tradition. *Down to Earth*, October 31, p. 56.

Sharma, C. B. S. R. (2003). "Temples, Tanks and Rehabilitation Programmes" [Communicated].

Venkataraman, R. and Sharma, C. B. S. R. (1983). "Environmental impacts on human settlements—Prospects for a study based on archaeological remains associated with Mangamma Salai," *Tamil Studies* 3: 64-70.

Von Oppen M. and Subba Rao, K. V. (1980). Tank irrigation in semi-arid tropical India, Part II: Technical features and economic performance, *Economics Progress Rep. no.* 8. ICRISAT, Patancheru, Andhra Pradesh, India, p. 52.

Yoganand, T. R. K., Amruth, M. and Sharma, C. B. S. R. (1993). "Ecosystem Approach to the management of traditional irrigation systems." [in] Congress on Traditional Sciences and Technologies of India. Section: Water Management. Indian Institute of Technology, Bombay 28 November to 3 December, 1993.

19

Sustainability of Irrigated Agriculture in India with Reference to Water Management

R. S. GOEL, V. B. PATEL and G. P. MATHUR

Introduction

The per capita availability of renewable fresh water in India fell from 5200 m^3 per annum in 1950 to 1860 m^3 in 2000. Deteriorating quality of surface and ground water is further affecting the net availability of water for consumptive uses. Safe water supplies and environmental sanitation are vital for protecting the environment, improving health and alleviating poverty. According to the World Bank estimates, annual costs of impact of water pollution on urban and rural health varies between US$ 83.44 billion and 30.76 billion out of the total annual costs of environmental degradation varying between US$ 137.58 billion and 6.72 billion. Thus, water pollution accounts for about 60 per cent of the major annual environmental costs in India.

The gross irrigation potential increased from about 22.6 mha in 1950-51 to 94.73 mha by 1999-2000. As a result, the food grain production had a fourfold increase from 51 million tonnes in 1950-51 to 206 million tonnes in 1999-2000. But there is a need for producing as much as 325-350 million tonnes of food grains by 2025 AD to meet the food requirement of the growing population. In the wake of the development of large scale irrigation projects, several serious problems of degradation of soils, water and environment have cropped up threatening the sustainability of agricultural production. Further, due to increases of the demand for water for domestic, industrial and recreational sectors, availability of fresh water for agricultural sector will continue to diminish in times to come. The quality of available water will also be reduced.

India requires a judicious mix of large, medium and small dams coupled with the integrated management of water available from all sources including ground water to meet our water, food, fibre, ecological and power needs. Power is the most important ingredient today for socio-economic development. The pollution caused by burning of fossil fuel to meet energy requirements is causing global concern. An economically viable option of renewable energy is through water resource projects; which will not only meet energy requirements but also ensure food security through irrigated agriculture and moderation of floods and droughts.

Availability of water will improve sanitation, safe hygiene practices and better health standards to make India a safer place for all its citizens.

Currently the population of India is little over 100 crores and it is expected that it is likely to reach a figure of about 150 crores by year 2050.

Agriculture is of fundamental importance in India's economy. It contributes 28 per cent to GDP (1995-96) and 67 per cent to employment (1990-91), while investments in agriculture amount to some 11 per cent of gross domestic investment. It generates some 10 per cent of total exports and following the economic liberalisation programme commenced in the early 1990s agricultural commodities are among India's fastest growing export sectors. Despite the tremendous economic development and growth of industries in India, the livelihood of about 70 per cent of our population still depends on agriculture, directly or indirectly. Over 170 million people engaged in agricultural practices avail the benefits of irrigation facilities.

Availability of land is expected to emerge as a major constraint on agricultural growth. Due to increasing demand of land for housing, rising level of urbanization and industrialization, increasingly larger quantity of agricultural land is being shifted to non-agricultural uses. In the past such loss of agricultural land was being compensated by converting forest land into agricultural land. However, there is no scope for further conversion of the forest land. The only possibility for fresh land is through reclamation of waste land and land affected by salinity, alkalinity and water logging. It is expected that such reclaimed land will compensate for the loss of agricultural land for non-agricultural uses. The Net Sown Area is, therefore, expected to remain more or less constant at the level of 142 million ha in future.

Given that the net sown area cannot and indeed should not be increased further, the availability of land can be augmented only through increasing cropping intensity. Enhancing cropping intensity will not only require expansion of irrigation facilities, but also higher efficiency of the irrigation system. At present the ultimate irrigation potential is estimated to be 113 million ha. With this irrigation potential and the current cropping intensity coefficient with respect to irrigation, the cropping intensity can go upto 149 as against the current level of 134.20. However, this will not be sufficient. Concerted efforts have to be made to increase the cropping intensity coefficient with respect to irrigation through water conservation measures and more efficient and judicious use of existing irrigation facilities.

Land and water both are scarce resources, indeed very scarce in our context. However, as long as it is not possible to irrigate all the land that requires irrigation, relatively speaking water is more scarce than land. Therefore, one of the major principles of allocation of water should be to maximize value from a unit of water rather than maximize value from a unit of land. The implication is that instead of irrigating each unit of land to its full absorptive capacity of water in a given command, it is preferable to expand the command of the irrigation project and distribute water thinly over the land. This principle would favour water economic systems of agriculture in most of the irrigation commands rather than water intensive systems. Given the weak pricing system, maximizing the yield from water would require a very careful intervention in water allocation and distribution. Water in this sense is not a private good, but it is social good.

Role of Irrigation in India

Irrigation has played a core role in agricultural production and growth. Although gross irrigated area (currently about 90 million ha includes double cropping) is about one-third of total cropped area in India; the production from irrigated lands greately exceeds the productivity of rainfed farming. In a monsoon dependent farming system with rains unreliable and large areas with rainfall less than 1,000 mm per annum, irrigation has been a crucial input for agricultural production. Irrigation enables a higher productive potential from the lands and significant production response from associated use of high yielding varieties, fertilizer and other inputs. Both at the national and regional levels, agricultural growth and rural development have closely followed the growth and regional pattern of irrigation expansion.

Impact of irrigation in India on production has occurred through four routes: cultivated area expansion, cropping pattern shifts, productivity enhancement and increase in reliability. Ascertaining irrigation's precise contribution is however difficult. This is in part because there are no official Indian statistical data summaries giving the breakdown of agricultural production under irrigated or rainfed conditions. Nevertheless, various estimates point to a contribution from irrigated agriculture to overall agricultural production of about two-third and/or higher. Even more difficult is the assessment of irrigated agriculture's contribution to agricultural growth. The average aggregate productivity of irrigated agriculture per unit of land would seem to be seven times that of rainfed agriculture.

With nearly all arable land now under cultivation and limited progress to date with major technological breakthroughs for rainfed crops, irrigated agriculture will likely remain the key contributor to India's agricultural development. Though it is of far reaching importance to enhance productivity from rainfed agriculture which occupies two-thirds of India's cropped area and for most rural communities is still the primary source of livelihood.

In terms of the relative share in India's created irrigation potential, surface water in canals accounts for 39 per cent, ground water (with a major private sector involvement) for 47 per cent and other water sources for 14 per cent. Over the 1951 to 1995 period, the publicly created and managed major and medium schemes increased three-fold while the privately created and managed groundwater segment increased sevenfold. Regionally, the major states in terms of irrigation area are Uttar Pradesh, Andhra Pradesh, Madhya Pradesh, Punjab, Rajasthan, Bihar, Haryana, Gujarat, Tamil Nadu, Karnataka, Maharashtra, Orissa and West Bengal. In most states, ground water irrigated area exceeds the surface irrigated area.

By far, the most important contribution of irrigation to Indian agriculture has been in facilitating the spread of the Green Revolution. The expansion of irrigation from the mid-1960s period the enabling environment for use of the new high yielding varieties and fertilizer, pushing the production frontier to a new height and transforming the rural sector. The most major gainers were in terms of both productivity and area share were wheat and rice. As a result, foodgrain production increased from 51 million tonnes in 1951 to 199 million tonnes by 1997; a compound annual growth rate of 3.0 per cent. Between 1980-81 and 1995-96, the average foodgrain yields nationwide increased from 1.0 ha to 1.5 ha. Irrigation expansion has also imparted a greater degree of stability to farm production.

India has now become the world's largest producer of milk and cashew nuts, second largest producers of rice, wheat, sugarcane, groundnut and vegetables, third largest producer of wheat, cotton, sorghum, potatoes, coffee, rapeseed-mustard and the sixth largest producer of maize. Irrigation is the major factor that made this possible.

National Commission for Integrated Water Resources Development Plan (NCIWRDP), 1999 has estimated that increase in irrigation facilities alone has contributed to about 52 per cent increase in food grains production and improved management, agricultural practices and high yielding varieties have contributed to 48 per cent increase in food grains production. Once water is available, crop lands begin to produce more and become less prone to droughts. Slowly, over time, animal based production also increases thus raising nutrition and health levels of people.

With an estimated two-thirds or more of agricultural production from irrigated agriculture, employment impact is clearly significant. The incremental employment generated by the irrigation potential created during the Eighth Plan has been estimated by Government at 8.7 million person-years. Since irrigation development forges inter-sectoral and inter-regional linkages through output growth and income flow, its employment and income impacts have powerful multiplier effects. A 100 rupee worth of irrigation induced incremental agricultural output is estimated to generate Rs. 105 worth of additional output in manufacturing and Rs. 114 in the tertiary service sector, implying an overall non-farm output multiplier of 2.19.

Irrigation has also played a role in poverty alleviation. Between 1964 and 1994, the national average prices of rice and wheat, the major Indian food staples, declined by 2.2 per cent and 3.3 per cent per annum in real terms, respectively. A significant inverse relation between the incidence of poverty and the extent of irrigation development has been found in several cross-sectional studies based on both districts and agro climatic sub-zones. While the incidence of poverty is as high as 69 per cent in districts with less than 10 per cent of cropped area under irrigation, it is about 26 per cent in districts where irrigation covers more than 50 per cent of cropped area and just 10 per cent in Punjab and Haryana with over 70 per cent of their cropped area under irrigation.

Environmental Impacts of Floods and Droughts

Floods

Over 40 million hectares of the area of the country experiences periodic floods. The average area affected by floods annually in India is about 7.5 m ha of which crop area affected is about 3.5 m ha. Floods have claimed on an average 1529 human lives and 94000 cattles ever year. Apart from loss of life and domestic property, the devastating effects of floods, sense of insecurity and fear in the minds of people living in the flood plains is enormous. The after effects of floods like the agony of survivors, spread of epidemics, non-availability of essential commodities and medicines and loss of their dwellings make floods most feared natural disaster being faced by human kind.

Crops grown in the flood plains suffer from congestion of water on the farmlands. Management of the surface water becomes a very tricky operation in the flood prone areas during periods of heavy rainfall. Floods also affect the vulnerable aquatic and wild life, forests, mangroves and precious bio-diversity in the flood plains. Large-scale damages to forests, crops and precious plants and deaths of aquatic and wildlife, migratory and native birds in various National Parks, Delta region, low altitude hilly areas and alluvial flood plains of Assam, Arunachal, Uttarakhand, U.P., Bihar, Orissa, West Bengal, have always been the matter of serious concern. River Valley Projects moderate the magnitudes as well as frequencies of floods. While some projects are specially designed to provide flood cushion in the reservoirs along with other benefits, others also help in reducing the magnitude of floods with proper operation and control of gates.

Droughts

It is estimated that around 263 million people live in drought prone area of about 108 m ha which works out to one-third of the total Indian geographical area. Thus, more than 26 per cent of total population of the country face the consequences of recurring droughts, on a wide spectrum of social concerns. River Valley Projects are designed to provide 'carry-over' storage in the reservoirs to help in mitigating the droughts.

During the drought years, there is a marked tendency of intensive exploitation of ground water, resulting in abnormal lowering of ground water table thus accentuating the distress. Grave adverse impacts are borne by flora, fauna and domestic cattle and the very life itself fights against nature for its survival. Droughts affect rural life in several ways. This accentuates problems in cities in the form of mushrooming of slums and pressure on the existing civil amenities thereby adversely affecting urban life.

Impacts of Water Storage Projects

There are numerous incidental benefits from the construction of large dams and canal systems such as improving environment and health, afforestation, fisheries development, tourism and recreational facilities, employment generation, development of agro-based industries, network of roads in command areas, development of land and improving general socio-economic standards of the people. It has been established through numerous case studies that significant improvement occurs in the food and nutritional level of the people with higher per capita food availability and diversification of crop production especially cash crops after introduction of irrigation in the area that are quite remunerative to the farmers. New employment opportunities generated by the intensification of agricultural and the associated economic activities further improve the financial condition of the people including landless labourers. As a multiplier effect, large river valley projects tremendously improve the health of rural population by significantly enhancing the education, health care, transportation facilities and the life styles particularly of women.

India is predominantly an agricultural country depending on monsoons during the four months of the year for which are to be used judiciously to cater to various needs for the balance period of the year. This warrants necessarily construction of reservoirs to store flows during monsoon for use during the other period to meet various demands. Water resource development requires a judicious mix of large, medium or small reservoirs, which are location specific.

Water resource projects may submerge forests, reduce downstream flows in rivers and at times lead to loss of bio-diversity. It may be noted that the loss of forest area due to submergence is less than five per cent of the total forest area lost in the country in the last five decades. The loss of biomass through submergence is, far smaller than the biomass generated on account of the irrigation. Notably, it has been observed that a forest far superior to the original, sans the original bio-diversity, comes up after the creation of the reservoir. Adverse effects like water logging and salinity are being prevented through conjunctive use of groundwater, prevention of canal water leakage, reduction of seepage losses from water carrying bodies, implementation of adequate drainage and adoption of efficient irrigation methods along with water conservation.

Reservoirs may create new conditions for the growth of organisms, and ultimately, as adjustments are made, foster new eco-systems. Varieties of new organisms thrive on this new eco-lake system. Additional water made available for the dry period of the year, when the environment tends to be harsh and makes the area inhospitable, supports the growth of life around. Water resources projects provide a dependable source of drinking water. People from the irrigated areas enjoy better health and sanitation facilities, thus reducing the incidences of diseases. General decline in incidences of diseases has been reported from different irrigation commands. The very availability of water leads to improvement in the level of sanitation. The improved economic status also makes people health conscious and capable of availing of requisite health care.

Substantial increase in the numbers of tigers, panthers, elephants and Cheetals have been observed in the famous Jim Corbett National Park with the availability of green fodder, clean water throughout the year and improved climatic conditions after construction of the Ramganga Multipurpose Dam Project. It is also observed that rare species of birds flock there after the reservoir constructed. Similar phenomenon of an increase in birds and wildlife has also been observed around the Rihand and Matatila reservoirs, which were previously barren lands. Some of the best tourist places of India like Ukai tourist resort, Periyar wildlife sanctuary, Shalimar garden, Brindavan garden, Pinjore garden, Kalindi-Kunj, Matatila garden, Dhyaneshwar Udyan and the Ramganga Udyan are the bye-products of river valley projects.

As per the UNICEF (1988) report, new water supply facilities sourced from large dams improved the sanitary conditions, which led to significant improvement in general health conditions. Water resources projects provide a dependable source for drinking water. People from irrigated areas in general enjoy better health and sanitation facilities, thus reducing the incidence of disease.

Vectoral risks can be reduced by removing sources of stagnant or slow-moving water and by ensuring continual maintenance of drains and canals.

The controversies concerning the rehabilitation of persons displaced by dams have muddied the entire debate on the utility of water resources projects and caused much harm to the national economy and well-being of the population at large. As per the broad assessment made by Central Water Commission through the review of data of 2784 dams spread in 14 States, the total affected persons may range between 6-7 million. The exaggerated claims by the opponents of large dams blow up this figure up to 70 million by taking the average of the recent few mega dams and multiplying the same by 4291. It has to be borne in mind that most of the high dams (by definition every dam having height of more than 15m is classified as high dam mainly for safety concerns) did not displace persons, firstly, due to very thin population in the submergence in earlier dams during construction, secondly, very few dams having the height greater than 50m would have the submergence impacts on the upstream habitation. Even though, the national policy for rehabilitation and resettlement of project affected persons is still to be enacted, liberal provisions and comprehensive plans for implementation are being kept in recent water resource projects so as to ensure that the PAPs are rehabilitated properly with adequate civic amenities so that their economic conditions improve after rehabilitation. Most of such PAPs reside in areas of extreme environmental fragility and largely deprived of nutritional food, potable water and productive employment.

Employment benefits of river valley projects have been widely experienced. These benefits first arise during the construction phase. Typically, 60 per cent of the capital costs of a major irrigation projects are in the shape of payment to construction workers. Against this one-shot employment benefit for the unemployed and the underemployed, sizeable recurring onfarm employment benefits are generated because labour use in irrigated farming is more than in unirrigated farming. Thus, irrigation development in a tract stems out migration of job seekers from that tract to distant centres.

Large dams help in conversion of wasteland into agricultural land and making the area greener. Indira Gandhi canal has not only transformed Western Rajasthan into vast green area but also checked the spread of Thar Desert in the adjoining areas of Punjab and Haryana. Bhakra Dam is a shining example, which has changed backward area of erstwhile-undivided Punjab into the granary of India with improved environment. The availability of water from Sardar Sarovar Project will benefit about 1.91 lakh people residing in 124 villages in arid and drought-prone border areas of Jalore and Barmer Districts of Rajasthan which have been suffering grave hardship on account of scarcity of water, besides checking the advancement of Thar Desert.

Need for Accelerated Water Storage Development

It must be noted that projected demands of water by 2025 can be met only by exploiting almost all the 1132 billion m^3 of utilizable water by conventional means. It would thus be essential to utilize the entire 690 billion m^3 of the utilizable flows for which storages of the order of 400 billion m^3 will have to be created. Considering the expected increase in demand, the storage capacities may have to be doubled by 2025. Domestic water supplies alone would require 52 billion m^3 of water. Water demand forecasts show that Rajasthan, Maharashtra, Gujarat, Haryana, Karnataka and Tamil Nadu could face heavy water supply shortfalls. The water shortages would be far more serious in the water defecit basins like the Cauvery, Pennar, Sabarmati, Krishna etc. To meet the increase in water requirements in the next 25 years, it would be necessary to ensure substantial augmentation of water supplies; requiring sufficient raising of water storage capacities, thus necessitating completion of new large water storage projects.

The recent Supreme Court judgement for Narmada Projects has also highlighted that against the utilisable storage of 690 cubic km of surface water resources of the 1869 cubic km of water resources storage capacity of all dams in India is only 174 cubic km, which is incidently less than the capacity of Kariba Dam in Zambia/Zimbabwe with the capacity of 180.6 cubic km and only 12 cubic km more than the Aswan High Dam of Egypt. The Supreme Court ruled that environment concern has not only to be of the area which is going to be submerged and its surrounding area; but the impact on environment should be seen in relation to the project as a whole. Thus, expansion of storage capacity by completing on-going projects and taking up new projects is imperative and is critically important, in the interest of enhancing the water availability.

Environmental Impacts of Irrigation

In terms of environmental quality *vis-à-vis* irrigation, it has different impacts on different areas in different conditions, with both gains and losses. Contribution of irrigation projects in agricultural production, employment, quality of life, drinking water, navigation, fisheries, micro climate, tourism etc. and the other is based on the performance of irrigation that it has created several problems to the environment.

Environmental impact assessment can be made at two stages: (i) at reservoir site, while constructing a dam and (ii) at the command area level. Essentially, it is an assessment of human induced environmental problems. It is because irrigation *per se* is not responsible for environmental degradation. The fact is that the problems are arising due to mismanagement of irrigation water and defective policies that encouraged its injudicious use.

The issue of policy concern and debate is the intensive irrigation in some restricted classes of land. Broadly, there are five land irrigability classes. Class I is highly suitable for irrigation with negligible obstacles, while Class II is suitable for irrigation with slight obstacles. Class III is moderately and Class IV is marginally suitable for irrigation with moderate to severe obstacles respectively. Land classified under Class V is completely unsuitable for irrigation with very severe obstacles in topography, drainage or soil. Unfortunately, with the introduction of irrigation in any class of land, the high water requirement crops are allocated larger acreage without any reservation. In fact, the recommended cropping pattern is ignored by the farmers in the race to increase profitability.

Broadly, the environmental problems in irrigated areas include: (i) water logging and salinity, (ii) over-exploitation and degradation of ground water, (iii) impacts on water quality mainly by building up of nitrate and pesticide residues, (iv) problem of weeds, (v) health hazards, (vi) watershed and catchment degradation, and (vii) siltation aspects.

Waterlogging

Irrigated agriculture not only raised food production and improved productivity but also brought in its wake many of the environmental problems. Waterlogging and soil salinisation in agricultural lands are the most serious environmental issues being attended on a global scale. The rise of watertable and salinisation of soils after the introduction of irrigation in the arid and semi-arid areas is an universal phenomenon. Rise in water table is through contribution of transported water to the ground water through seepage from main canals, distributaries and field channels, deep percolation of water applied in the fields due to mismanagement in various ways and in appropriate cropping patterns. The developmental activities like the canal network and roads and railway lines in the irrigation project area also obstruct the natural drainage of rainwater causing accumulation of rainwater for longer periods and consequent greater recharge to the ground water. The problems of waterlogging and soil salinity in India were noticed even in the last century.

Waterlogging refers to the conditions when ground water is at or near the ground surface at least for some time in a year. According to National Commission on Agriculture an area is termed as waterlogged when the watertable attains a level at which the soil pores in the root zone become saturated resulting in restriction of the normal circulation of air, decline in the level of oxygen and increase in the level of carbon dioxide.

Shallow watertable may not be a curse, particularly when it is free from salinity hazards as happens in most humid regions. This contributes substantially towards the crop evapotranspiration, depending on the depth and the capillary rise which is a function of soil texture and porosity. However, not all crop varieties can tolerate very shallow depth of watertable owing to oxygen stress. The waterlogging also creates conditions for surface water stagnation and excess moisture after an irrigation or a rainfall. The prevailing wet conditions affect land preparation and sowing and also the inter-cultivation practices. The ill effects of waterlogging are manifold in areas which have saline soils and saline ground water.

Since it is difficult to measure the direct effects of waterlogging in the field, these are related indirectly to the depth of watertable. The soil type may influence the performance of the crop under excess moisture conditions. As a result of the studies conducted so far within or outside the country, the required depths of the watertable for different types of soils and crops are generalised.

The watertable fluctuates with season attaining the lowest depth before the onset of monsoon and subsequently rising to the shallow level in the post-monsoon period. A 3 m watertable depth or more before the onset of monsoon is quite desirable to accommodate the percolated rainwater, leaving one metre depth for root growth. Thus areas having watertable within 3 m can be considered as critically waterlogged, demanding immediate implementation of drainage practices to lower the watertable.

Desirable Minimum Average Watertable Depth

(m below soil surface)

Crop	Soil		
	Sandy	Loam/Silt	Clay
Grasses	0.5	0.6	0.7
Cereals	0.6	0.7	0.8
Crops grown for fodder	0.7	0.8	0.9
Tuber and root crops, fibre crops, oilseeds, sugarcane, vegetables	0.8	0.9	1.0
Orchards and trees	1.0	1.2	1.4
Temporarily dry or fallow land with upward seepage of saline ground water	1.2	1.5	1.2

Areas having watertable within 3 to 6 m can be termed as sub-critical and necessary drainage and water management are to be planned for the immediate future. In areas where the prevailing watertable is between 6 and 10 m, precautionary measures such as onfarm water management practices, better water conveyance system, conjunctive use of ground water and surface waters and improvement in surface drainage system need to be undertaken.

Soil Salinity

The rise of watertable in semi-arid and arid areas may mobilise the salts present in the soil profile and groundwater. Initially the salts present in the soil profile are leached down and the soil in the root zone becomes salt free enhancing the agricultural production. The rise of watertable may disturb the hydraulic gradients in the area and may induce movement of saline ground water into non-saline areas and into streams creating stream salinity problems. Once the ground watertable rises within 2-3 m of the soil surface, it contributes substantially to evaporation from the soil surface and water uptake by plants. The upward flux of water due to evaporation and water uptake by plants results in gradual concentration of salts in the shallow depths of soil profile. In the initial years, the crop yield may be reduced due to salinity, but as the severity of the problem increases, the lands may have to be abandoned since the excess salts render the soils completely unproductive.

Soil salinity problems develop whenever soil and hydrological conditions favour the accumulation of soluble salts in the root zone. The common sources of salt are salts present in the soil and ground water and salts applied through irrigation. The source of salts in the soil could be that transported along with the alluvium and or that accumulated from weathering of the soils. These soils generally do not occur extensively in the humid areas where the salts get continuously leached out of the soil by excessive rainwater. Saline soils, however, are quite common in the semi-arid and arid regions receiving inadequate precipitation to accomplish leaching of salts originally present in the soil profile.

Soils of the arid and semi-arid areas generally contain salts at some depth. The depth of the salt layer would depend upon the water transmitting characteristics of the soils and the climatic conditions, particularly the rainfall. In the black soil area because of higher clay content predominantly of mont-morillonite, the effective rainfall is much less than the actual rainfall due to higher run off. Hence, there is no scope for leaching away of the salts generated due to weathering.

Over Exploitation of Ground Water

Over exploitation of ground water is also a serious problem in some of the regions endowed with good quality water. Electric pumps and tubewells find extensive use to exploit this vital resource. Since there is no state regulation for the use of this common property resource, the ground water levels are drastically falling in most intensively cropped areas of the country as withdrawal exceed the recharges. Such a phenomenon is witnessed particularly in Punjab, Haryana, a few pockets of Western Uttar Pradesh, Tamil Nadu and Western parts of Gujarat. The following are the major reasons for the excessive use of ground water: (i) Inadequate availability on absence of canal water, (ii) no private cost of extracting additional quantity of water, (iii) huge subsidies or flat rates of electricity to extract ground water, and (iv) low rainfall in the region. The water table in Karnal district of Haryana fell from 4.8 metres in 1974 to 7.7 metres in 1989. The corresponding figures for district Kurukshetra were 7 and 10.7 metres. The water table in Punjab reported to recede at a rate of 0.3 to 0.5 metre per year. Thanks to the canal seepage in these areas that has recharged for ground water otherwise the falling rate was expected much faster. Cross district analysis showed that rice acreage and number of tubewells were responsible for overexploitation of ground water.

The implication of fast decline in water table are: (i) increase in pumping cost over time, (ii) widening the disparity between small and large farms, and (iii) formation of poisonous gases in the wells resulting into deaths of human beings.

There is a need to optimally use ground water and wherever possible link its use with availability of canal water, seepage losses, rainfall etc. Any price policy or technical strategy of utilizing ground water in isolation may be a futile exercise.

Ground Water Degradation

The presence of naturally occurring traces of minerals, e.g. fluoride, arsenic, iron, is a source of groundwater contamination. However, toxicity of ground water has been increased further by the leaching of agricultural as well as industrial chemicals into soils. In states such as Gujarat, groundwater has been degraded by increase in nitrate concentrations due to application of fertilizer to cultivate cash crops, including oil seeds, cuminm, sugarcane, cotton and groundnut. Land disposal of untreated wastewater and hazardous wastes from industries poses a severe threat to groundwater quality. Heavy metal contamination is particularly common in thermal power, tannery and mining intensive areas. Additionally, excessive withdrawals of groundwater, coupled with decreased recharge rates, have resulted in contamination from salt water intrusion in coastal areas. Aquifer degradation is a particular problem in states such as Gujarat, West Bengal and Tamil Nadu.

Impacts on Water Quality

The preservation of water quality has direct implications for water supply as water quality determines effective water availability. Similarly, a minimum quantity of water is required to maintain the assimilative capacity of a water body to receive effluents and/or guard against contamination (e.g., saline intrusion). The inter-relationship between water quality and quantity is clearly demonstrated in large cities such as Delhi where limited water supply is often unsafe for consumption. Environmental and health issues are, therefore, integral to both inter-sectoral and inter-state issues.

Agricultural Pollution: Agro chemicals—fertilizers and pesticides—are the main contributors to surface water pollution from agricultural areas. Although the average rate of fertilizer per ha is relatively low in India. The use of agro chemicals is unevenly spread throughout the country. In states where consumption is high such as Punjab, Tamil Nadu, Andhra Pradesh and Haryana, eutrophication of lakes, reservoirs and tanks is an environmental concern. Additionally, a number of pesticides classified by the World Health Organisation (WHO) as highly hazardous that are either banned or highly controlled in Western countries are commonly used in Indian Agriculture. The impact is far more permanent on groundwater resources. Irrigation has induced to use more fertilizer and pesticide. Fertilizer consumption increased because it complimented with irrigation in augmenting production. While pesticides use has gone up due to contrtol of more diseases and pests.

Impact of Fertilizer

The important problem related to fertilizer *vis-à-vis* environmental quality is nitrate pollution of ground water, besides eutrophication of lake and river water, increased emission of gaseous nitrogen and metal toxicities. There are reports from developed countries that fertilizer related pollution is increasing rapidly and affecting agricultural production and deteriorating the quality of land and water. There are reports of nitrate pollution of ground water and surface flows through saline seeps originating from feed lots or high fertilizer fields. The available evidences showed that with most efficient agronomic practices, fertilizer N use by upland crops seldom exceeds 50 per cent. It has been estimated that India is losing roughly 2.5 million tonnes of fertilizer N per year out of a consumption of 7.2 million tonnes. In the absence of sound agronomic and management practices, country may stand to lose approximately 6.7 million tonnes by the end of this century. Nitrates taken in with water or food result in gastric concert and methaemoglobinaemia, commonly known in infants as the blue-baby syndrome.

The present evidences under Indian context reveal that nitrate-nitrogen does not foresee a serious problem to Indian agriculture. However, government priority and urgency to increase agricultural production in years to come to meet the demand for food grains and huge subsidy on fertilizer to encourage its use could lead to escape of a part of applied nutrient into waters of highly fertilized areas. A strict monitoring of the quality of ground water is thus necessary.

Impact of Pesticides

Indian agriculture is using relatively less pesticides as compared to developed countries. However, its use overtime has increased from

2000 tonnes in 1950s to 80,000 tonnes in 1980s and 120,000 tonnes in 1989-90. Agriculture sector uses two-third of total pesticides use in the country and five states, namely Punjab, Gujarat, Maharashtra, Tamil Nadu and Andhra Pradesh use about 90 per cent of the total pesticides.

Injudicious use of chemicals in agriculture deteriorates land quality, contaminate water, food and environment. The spectacular spurt in pesticide use has resulted in secondary pest outbreak. The pesticide residues in soil may create a variety of hazards. Soil micro-organism which cause breakdown on cellulose, nitrification, turnover of organic matter and other biological materials may be adversely influenced by pesticides. There are reports that pesticides and chemical inhibit the microbial population in soil thereby resulting in reduced nitrogen fixation by symbiotic bacteria. Several evidences are reported on serious decline in population of earthworm due to pesticide residues and affecting the yield of different crops.

The toxic affects of pesticides and their residual affect in food and later may affect human and animal health; eventually their productivity. Continuous use and exposure of these chemicals lead to certain diseases among farm labourers and farmers employed for spraying operations.

Problem of Weeds

The problem of weed infestation is very common in canal irrigation projects because of adequate and reliable supplies of water, light and nutrients provide optimal conditions for their growth. The magnitude of aquatic weeds varies from one project to another. A survey was conducted by the National Committee on Environment Planning and Coordination to assess the magnitude of weed problems in different parts of the country arising from canal irrigation. The survey report that water hyacinth, cattails, water lettuce and water fern were the most threatening weeds, especially in the States of West Bengal, Bihar, Orissa and Assam. These weeds infested roughly 40 per cent of the total cultivated area (8,00,000 ha) of these States. Among these weeds, water hyacinth was most severe, covering an area of about 1,50,000 ha in West Bengal and 50,000 ha elsewhere in the country. Similarly *Salvinia Sp* has already spread in 60 per cent of the total water surface of the Kakki Reservoir in Kerala within a short span of four years. These weeds adversely affect the crop growth and yield. Besides affecting the crop growth, the aquatic weeds pose a threat to irrigation and drainage systems, fisheries, navigation etc. These also create a favourable environment for growth of pests and vectors of several epiderminological diseases.

Weeds complete with young seedlings for moisture and nutrients. Maintaining a weed free field for the first 30-40 days is crucial in sustaining optimum crop stands. Experimental evidence confirms that if the seeds are unchecked the yield of crops can decrease upto 50 per cent or more. To lessen the burden on labour employment for weed control, integrated weed management practices using chemicals, cultural practices and manual weeding devices have been developed. Simple weeding devices (single row dryland weeder) should be an important part of the drive to eradicate weeds.

Health Hazards

Spread of some diseases to human health is another adjunct of canal irrigation. The major water borne diseases observed to be very severe in the rural areas are dysentery, cholera, malaria and filariasis. The Government of India spends huge amounts of money on disease eradication and health improvement programmes in the rural areas still these water borne diseases keep on increasing over the years unabatedly. Punjab, Haryana, Andhra Pradesh and Uttar Pradesh have now become endemic for malaria on account of the higher water table, waterlogging and seepage in the canal catchment area. There are evidences from Raichur district of Karnataka that after the construction of Tungabhadra Dam and canal from Raichur district of Karnataka that after the construction of Tungabhadra Dam and Canal network, malaria became highly enendemic. Reports of increasing incidence of malaria and filariasis have also been reported in the Saryu Canal and Gandak Canal command areas.

Watershed and Catchment Degradation

Weakness in land management are inextricably linked with India's water resource management problem. Mounting population pressures with still insufficient techniques applied for watershed, land and water resource management have caused over-grazing and loss of forest land. About 6,000 million tonnes of top soil are lost annually in India through erosion. Not only are soils degrading but also the retentive capacity of the land is reduced by lack of vegetative cover. Flash flood run off is thus exacerbated rather than the more beneficial percolation of rainfall as ground water. This further limits cropping possibilities on residual moisture and exacerbates ground water depletion, including drying up of traditional sources of rural drinking and domestic water. Catchment management and local level micro catchment management techniques already known and promoted in India need more widespread adoption.

Siltation Aspects

Part of the top soil losses flow into the streams and rivers causing changes in their hydraulic regime. Siltation is a problem in surface water canals and India's many reservoirs and tanks. This adds to maintenance costs in canal and drainage networks and for the reservoirs and tanks, significant loss of storage capacity at an annual rate estimated to range from 0.5 to 1.2 per cent. Losses in storage translate into reductions in hydroelectric power production, inadequate irrigation water storage and reduced protection.

Suggested Strategies

The causes of human induced environmental degradation are different in the irrigation command areas, their remedial measures are also different. On the biological side, the presence of bacteria, algae and protozoa very much depends on the basic physio-chemical status of water. Of particular importance is the presence of oxygen in water. That is why the biological oxygen demand in ml/litre is used as an important measure of the concentration of the organic pollutants. Addition of nutrients to natural water results in the growth of algae and other aquatic plants. This is a natural process, but it gets considerably accelerated when nutrients are introduced through city effluents. Eutrophication can have a significant effect on domestic, industrial and recreational use of water.

While dealing with water, as distinct from the land and air, we have to go beyond the concept of environmental protection and adopt the strategy of environmental enhancement. Water is the dominant element in the environment which can bring about a desired change. Land and air are not easily manipulable like water. Hence water has an additional role to play, namely that of improving the environment.

For example, organised and efficient distribution of canal water to Sardar Sarovar Project (SSP) over a vast command area (34,000 sq. km) to about one million individual farm holdings creates challenges for the management and economic viability of the infrastructure, apart from environmental sustainability. The SSP is planned with extensive irrigation formulation wherein timely availability of limited water is of fundamental importance. To accomplish the onerous and complicated task of fundamental importance, the entire irrigation network has been planned, with village areas, as independent units known as "Village Service Areas" (VSA). Broadly, better on farm management, provision of agricultural land drainage, conjunctive use of surface and ground water, environmental friendly input output pricing policy and formation of Water Users' Associations (WUA) may encourage judicious use of irrigation water for sustained agricultural development.

Macro planning or inter-sectoral policies in regard to water development currently favours a mix of small and large projects, surface water and ground water development, lift irrigation and gravity irrigation etc. Irrigation schemes tend to be location specific and in most cases alternatives do not exist. Hence this policy of a judicious mix will have to continue. Where option exists, this would have to be evaluated. However, conjunctive use of ground water may have to be insisted upon for surface irrigation schemes as a result of sustainability. In regard to project planning, the recent policy already includes provision for additional live storage to allow irrigation benefits to continue undiminished for at least 50 years. A stricter and more meticulous planning of the conjunctive use at project planning stage seems essential. Central Water Commission has already formulated a draft policy in this regard.

In regard to project design, canal design practices based on the regime theory design of channels has proved successful in terms of sustainability. However, allowing partial discharge, particularly in the secondary channels seems to create considerable hydraulic problems and this practice may have to be changed. The present policy does not favour canal lining except for special problematic reaches and this seems to be a sound practice. Planning policy changes which favour shift to high technology through drip and sprinkler irrigation. Provision of "Offline" storages for maintaining supplies in the inter-rotational period to be incorporated in future designs.

In regard to the complex problem of operation and maintenance and financial sustainability, policy changes favouring increased water prices and turn over of tertiary water distribution to water users associations have been incorporated in the new irrigation management policy.

The policies may be aimed to direct a large part of subsidy from irrigation sector to the preventive measures of soil degradation to restore the production potential. Incentives must be extended to encourage the adoption of water saving devices. A blend of technical, economic, social and political aspects should form a part of developmental strategy to achieve the twin objectives of increasing production and improving environment.

Selection of Crops and their Varieties

In areas irrigated with brackish water, the crops which have lower water requirement are sown. There is wide variation in tolerance of the varieties of the same crop to saline or alkaline environments. Oilseeds are more tolerant to saline water than pulses.

In areas with highly saline water, grasses like Rhodes grass, Bermuda grass, Kamal grass, alfa alfa, rype and oats may be grown for fodder purposes. With appropriate management practices poor quality water can be used with minimum hazard for sustainable crop production in the given agroclimatic environment. Limits of salinity of irrigation water for use in agriculture and important cereals, oilseeds, pulses, legumes and vegetable crops have been identified under the All India Coordinated Project on use of saline water for agriculture and published in the CSSRI technical Bulletin No.19 (1994).

Conjuctive and Cyclic Use of Water

When multi-quality water from various sources, viz., canals, tanks, tubewells and dugwells is available, water can be used for irrigation of crops by either blending or cyclic use (alternate use of good and poor quality water). For blending or conjunctive use, water from canals, tanks etc. should be available in adequate quantity at appropriate times for mixing with poor quality water.

The other alternative is to use good quality water from tanks, canals and other surface sources which are not adequate for the full irrigation requirement of crops in more sensitive stages like seed germination, seeding establishment, flowering etc. Saline water may be used for irrigation once the crops are well established and the crop attains a more tolerant stage. This cyclic method of irrigation with alternative use of good quality and poor quality water is being followed by farmers in some states. Experiments conducted at CSSRI, Karnal have proved that one pre-sowing irrigation with canal water and four subsequent irrigations with saline drainage water resulted in good yields of wheat and most of the salts accumulated due to irrigation were leached down in the rainy season. Drip irrigation for high valued crops like orchards and some vegetables as done in Isreal may also be feasible in our country.

Reclamation of Coastal Saline Land and Prevention of Salinity Ingress

No recharge or salinity control scheme will give its full long term advantage unless supported by proper management techniques. This can be achieved by proper control over the use of ground water, conjuctive use of ground and surface waters and lower water requirements.

Some form of artificial recharge techniques has to be adopted in areas where ground water withdrawal is higher than the natural recharge. Artificial recharge can be undertaken by creating storage of fresh water and maintaining the head for as long a period as possible to accelerate the induced infiltration. It can also be accelerated by effecting increase in the detention time of the surface runoff by providing vegetational cover and afforestation. Check dams, recharge tanks, spreading channels and recharge wells are the measures using storage on existing rivers. Their effect will depend on the type of geological formation which are under water.

Seawater ingress takes place through aquifers out cropping in the sea and also through tidal water flow into river channels. Tidal regulators and *bandharas* prevent upland movement of tidal water. The subsurface ingress can be prevented by freshwater, static or extraction barriers. Tidal regulators have got a control device with gateways, while *bandharas* have the crest at or above the high tidal level. A fresh water barrier is used for creating a fresh water head balancing the seawater head. This is suitable when fresh water is available and aquifers are homogenous. A static barrier though costly, can be adopted where there is no surplus water. An extraction barrier requires costly maintenance although it is a good device and may be adopted in developed areas where civil works are not possible.

Role of Communities, Women and Professional Societies

It is essential that environmental aspects and the process of planning and operation of water resource projects be fairly understood by the experts of different disciplines. Participation of people is a must in the management of water. People have to be made an integral part of the water management system. The community is to be made not only water conscious, but also to be integrated to participate in the planning and management of such projects and pollution prevention programmes. It is unfortunate that a smear campaign has been launched during last two decades against hydropower and water resource projects by exaggerating the likely or assumed adverse environmental impacts and by suppressing their need and tremendous benefits.

Gender issues in integrated water management especially the role of women needs to be underlined. On an average, the rural women in India make around 9 trips to a water source, collecting 192 litres of water for an average 6-member household. Implementation of water resources projects brings relief to the women by providing water nearer to the homesteads so that they need not traverse long distances on foot to fetch water. There is also a need to involve women in a big way in all the activities of integrated water management such as participatory management, decision-making, process, technology improvement, knowledge, information and training programmes.

The presentation of facts, analyzed inferences and voicing of unscientific perceptions have to be delinked. The professional societies can perform like watchdogs in this respect. Professional societies can act as multi-disciplinary fora for national and regional debates, analysis and framing of action plans on water related matters by utilizing their infrastructure, professional expertise, library, publication and documentation services. These societies can serve as rich sources in generating technically sound options with well-defined limitations and assumptions in Indian peculiar situations for taking informed decisions. The services of professional Societies like Institution of Engineers (India), Indian National Science Academy, Indian Water Resources Society, Indian Association of Hydrologists, Indian Society for Hydraulics, Water Management Forum, Indian Society for Water Management, Rural Development Forum having vast network, good spread and pool of expertise may be channelised in debating, dissemination and documentation of reports and action plans for integrated water and land management for long-term sustenance.

Conclusion

The environmental impacts of irrigation management and water storage projects have been discussed in this article. The importance of food security and role of irrigation have also been reviewed in Indian context. Non-development of water storage projects is not a viable or available option in India. Accelerated development of hydropower is highly desirable which would not only help in meeting our energy requirements but will also ensure food security due to increased irrigation from the regulated flows.

The issues of water quality ranging form agro-chemical, industrial and domestic pollution; impacts of water use on ground water levels (including depletion or conversely water logging and salinity); soil erosion and siltation; degradation of wetlands; impacts on flora and fauna and various health-related problems are becoming a matter of crucial importance. Many of these problems are less evident to the general population than the more visible quantity related issues. Ground water related issues for instance are gradual in impact. Rising water tables in Northwest India due to over a hundred years of irrigation have only in some localities reached the root zone, yet if this trend continues without appropriate drainage, large tracts of currently irrigated lands will become salinised and go out of production.

REFERENCES

Chandra, Suresh and Goel, R. S. (2001). "Environmental Impacts of Hydropower and River Valley Projects and Quantum Need of Large Storage Projects in India", *CBIP Seminar on Accelerated Hydropower Development and Pumped Storage*, Chandigarh, 9th-10th August, 2001.

Dhawan, B. D. (1999). *Studies in Indian Irrigation*, Commonwealth Publishers, New Delhi.

Goel, R. S. (Ed.) (1993). *Environmental Impacts of Water Resources Development*, M/s Tata McGraw Hill Publishing Company, New Delhi.

Goel, R. S. (Ed.) (2000). '*Environment Impacts Assessment of Water Resources Projects—Concerns, Policy Issues, Perceptions and Scientific Analysis*' M/s Oxford & IBH Publishing Co. Pvt. Ltd., New Delhi.

Goel, R. S. (Ed.) (2000). "*Environmental Management in Hydropower and River Valley Projects—Techniques of Management, Policy Issues, Case Studies and*

Application of Scientific Tools, M/s Oxford & IBH Publishing Ltd., New Delhi.

Goel, R. S. and Srivastava, R. N. (Ed.) (2000). *Hydropower and River Valley Development-Environment Management, Case Studies and Policy Issues*", M/s Oxford & IBH Publishing Co. Ltd., New Delhi.

Joshi, P. K.,(1912) "Environmental Impacts Assessment of Irrigation Projects", Seminar on Irrigation Water Management, *Water Management Forum*, New Delhi.

Ministry of Water Resources, (1999). *Report of the National Commission for Integrated Water Resources Development*, New Delhi.

Ministry of Water Resources, (2001). "*Report of the Working Group on Water Related Ecological Matters for X Five Year Plan*", New Delhi.

Prasad, Kamta and Goel, R. S. (Ed.) (2000). *Environmental Management in Hydro Electric Projects*, M/s Concept Publishing Company, New Delhi.

20

GIS-Based Study for Drinking Water Supply

SHAILZA VERMA, RENU BHARGAVA
and P. K. GARG

Introduction

The physical environment mainly consists of landforms, climate, soil, water, minerals, and natural vegetation. This resource complex is necessary for human existence since it provides human-beings with living space, food and a variety of raw material for fulfilling their basic needs. There will be more pressure on natural resources, to meet the demands of growing population. Economic considerations, ignoring ecological consequences, has placed existing natural resources in a critical condition. Environmental engineers are concerned to protecting the environment from the potentially deleterious effects of human activity, protecting human populations from the effects of adverse environmental factors and improving environmental quality for human health and wealth being. If the management strategy for natural resources is left as it is, it is predicted that natural resources in the near future will no longer meet the demand.

Water is the most important component of life on the Earth. It is required for domestic purpose, irrigation, industries etc. Environmental problems occur, when either input is less or demand is beyond the regenerative capacity of the sources. The availability of water supply in terms of both quality and quantity is essential to human existence. In recent years, demand for drinking water has increased manifold. High rate of population growth further compounds the problem. Therefore, in order to ensure the availability of sufficient quantity of good quality water, it becomes essential to plan and build sustainable water supply schemes, which may provide potable water according to demands. Identification of sources of water supply, their conservation, and optimal utilization is of utmost importance (Mudgal, 1999).

The work in this paper focuses on the drinking water supply problem in rural areas, and suggest the measures through source selection and utilization to minimize the problem of drinking water. The work has been carried out in GIS environments, for easy handling and analysis of the data needed for the water supply.

Selection of Water Supply Sources

The objective of a protected water supply system is to supply safe and clean water in adequate quantity, conveniently and as economically as possible. The following points are usually considered in selecting a particular source of water supply—

- *The Quantity*—The quantity of water available at the source should be sufficient to meet water demands taking into account the daily and seasonal variations and projected growth in the size of community being served. The demand of water is evaluated according to the present population and supply is evaluated by the quantity of water available.
- *The Quality*—The quality of water should confirm to the standards and should be potable. Impurities should be limited, so that with appropriate and feasible treatment, water can be supplied to the community.
- *Protection and Feasibility*—Watershed should be protected from the pollution with human excreta. Sources available should be such that water can be applied at reasonable cost (both in absolute term and in comparison to possible alternative source).
- *Elevation*—Sources should be as near as possible to the habitation. It is preferable to have a source situated at an elevation, higher than that of a village to be supplied with water flowing through gravity.

Role of GIS

Remote Sensing and GIS plays an important role for providing water related information and GIS helps in carrying out modelling and analysis for sustainable development of water resources. These technologies may save considerable amount of time and money for such studies. Environmental applications of GIS include, site assessment and cleanup, wildlife management, pollution monitoring, risk analysis, vegetation mapping and public information.

GIS is a computer system, capable of assembling, storing, manipulating and displaying geographically referenced information, i.e., data identified according to their locations. This technology can be used for scientific investigations, resource management, and development planning. GIS is a set computer programmed that manipulate and analyze spatial data. It represents one of the hottest growth areas in the world of computing. This tool has made the data handling and analysis much easier. In addition, it has the advantage of handling attribute data, in conjunction with the special features which were earlier not possible with the manual cartographic analysis. GIS technology, data structures, and analytical techniques are gradually being incorporated into a wide range of management and decision-making operations.

Study Area

Dehradun district is situated in foothills of Himalayas. It is divided into 6 blocks—Vikasnagar, Chakrata, Doiwala, Sahaspur, Raipur and Kalsi. Sahaspur block is situated in the central part of Dehradun district. Latitude and longitude of the area is 77° 46' to 78° 10' and 30° 14′ to 30° 29′. The total area of Sahaspur block is 38 sq.kms. The Sahaspur block has 119 villages, in which seven villages, viz. Chaknaugaon, Bishanpur, Chaksaliawala, Gopiwala, Shahpur Santour, Malukawala and Khamadoz and Manduwala are uninhabited according to 1991 census data. Total population as per 2001 census is 102749.

Sahaspur block is basically rural area of the Dehradun district, which contains both the hilly area and plane area. In the north Mussoorie hills are situated, which is the glamorous part of this block. In this block industrialization and forest storage resources centre cause not only the environmental problem but transportation problem also. Other environmental problems are created because of conversion of agriculture and fertile lands for urbanization. Low productive agriculture due to mining activities near the Mussoorie hills and landslides and depositions of boulders on agricultural lands. Education and medical facilities are also inaccessible. It has Jhajra and Malhan forest range in the south and Mussoorie forest range area in north. Main fruit in Sahaspur block is mango and litchi (NEERI, 1994).

Status of Water Supply Sources

In the rural areas, pipe water supply is done for 30 per cent of the population and 70 per cent of the population is served by tube wells, hand pumps, open wells etc. Average designed demand is 40 lpcd. Main sources of water in Sahaspur block are:

1. *Tap Water*—According to the Draft Perspective Plan Report,1992-2001, pipe water supply is not available in five villages; two of them are uninhabited villages. Other villages used rivulet water or other sources. Piped water supply is sufficient to fulfil 30 per cent of total demand.
2. *Mark II Hand Pumps*—The depth of India Mark II hand pump is (21-45 m), more than the private hand pump. A few shallow hand pumps have the problem of rusting. The sulphate concentration is also higher in private hand pump. A wide range of soil strata and soil characteristics present may also affect the ground water. Total number of Mark II hand pump in Sahaspur block is 49. Total discharge of Mark II Hand pump is 12 lit/day for 20 hours in a day.
3. *Tube Wells*—A total of 28 tube wells are present in Sahaspur block according to the Irrigation Department, Tube-well Division, Dehradun. Each tube well is used to irrigate 100 ha area in downstream direction.
4. *Open Wells and Shallow Hand Pumps*—They are present in the lower part of Sahaspur block, where ground water potential zone is also very good.
5. *Springs and Rivulet*—They are found in upper part of Sahaspur block. In the hilly areas, only springs and surface water are the main sources of water.

Methodology

Data Collection—Data for this study were collected from different agencies/offices and publications. In the present study, several base maps have been prepared using GIS which are essential for planning the work. Non-spatial data were also collected, as given below:

1. Population of all the villages of Sahaspur block—collected from Statistical Department, Dehradun.
2. Water supply schemes operational in the area.
3. Discharge from tubewells and hand pumps.

Digitization—For the present work, village boundary map for Sahaspur block has been digitized using ARC/INFO software. The errors have been corrected through ARC/EDIT software. After each editing operations 'clean' and 'build' commands were used in ARC/INFO environment to construct topology and create feature attribute tables.

1. *The Village Boundary Map*—The village boundary map was prepared from block level maps prepared by Lal Bahadur Shastry Academy of Administration, Mussoorie. Each village has been assigned with a unique code. In Sahaspur block, there are 122 villages, Jhajra forest range, Malhan forest range and Mussoorie forest range. This map has been taken as base map to create all other themes on it.
2. *Population Data*—Census data of villages for year 1991 were taken from Statistical Department Dehradun. Since current population data of Sahaspur block was not available, a population estimate has been made by taking population growth of 10 per cent per decade from 1992 to 2001.Two separate coverages were prepared for the population details of the area, i.e. for year 1991 and 2001.
3. *Tube Wells*—It shows locations of all the villages having tubewells. The data was collected from the Irrigation Department, Tubewell Division, Dehradun.
4. *Mark II Hand Pumps*—It shows the locations of all the villages having Mark II hand pumps. This map is prepared by collecting data from Garhwal Jal Sansthan, Dehradun.
5. *Surface Water and Springs*—This shows the locations of villages

using rivulet and spring water. The relevant data has been collected from the Garhwal Jal Sansthan, Dehradun.

6. *Slope*—Slope map has been generated from the contours taken from Survey of India toposheets at 1:250,000 scale (sheet nos. 53E, 53F, 53J, and 53K). A Digital Elevation Model (DEM) was created for this purpose. The area is divided into three zones in terms of slope percentage as 'moderate', 'steep' and 'very steep'.
7. *Rainfall*—Rainfall map is derived from Dehradun District Planning Map Series, Survey of India. The rainfall intensity is represented by two categories, as high (>1650 mm/yr) and low (900-1650 mm/yr).
8. *Soil Map*: The soil thematic map has been prepared from 1:1,000,000 scale Dehradun District Planning Map Series. There are two types of soils in the Block, viz., Brown Hill soil and Bhabar soil.
9. *Geology Map*—Geology map (Fig. 20.1) is compiled from 1:1,000,000 scale geological map of western Himalaya, prepared by K. S. Wadia Institute of Himalayan Geology, Dehradun. Geological composition, *old doon gravel* consists of boulders and clays and *young doon gravel* consists of gravels and sands.
10. *Ground Water Potential*—It is taken from a study carried out at IIT Roorkee, by Pulla Reddy (2002). Tables 20.1 shows area covered by various categories of ground water potential.

Table 20.1: Ground Water Potential Classes

Sl. No.	*Ground Water Potential*	*Area (km^2)*
1.	Very good	160.991
2.	Good	103.323
3.	Medium	75.213
4.	Poor	50.235
	Total	389.762

Identification of Villages Having Shortage of Water—Using all the available water sources maps, a composite map has been prepared which is shown in Fig. 20.1. It indicates total available water sources for each village. It is observed that in the upper portion of the Sahaspur block at altitude higher than 3000 m, only spring water and rivulets are available as sources of drinking water. In lower portion of Sahaspur block, tube wells and India Mark II hand pumps are available. In some villages, such as Bhagwanpur and Dungal Gaon, not even a single source of water is available. Some other villages do not have tap water at all. Tap water supply caters only 30 per cent of demand. The villages with piped water supply also have shortage of water, as the supply is not regular throughout the day. In the area, 10 per cent supply is done through other sources.

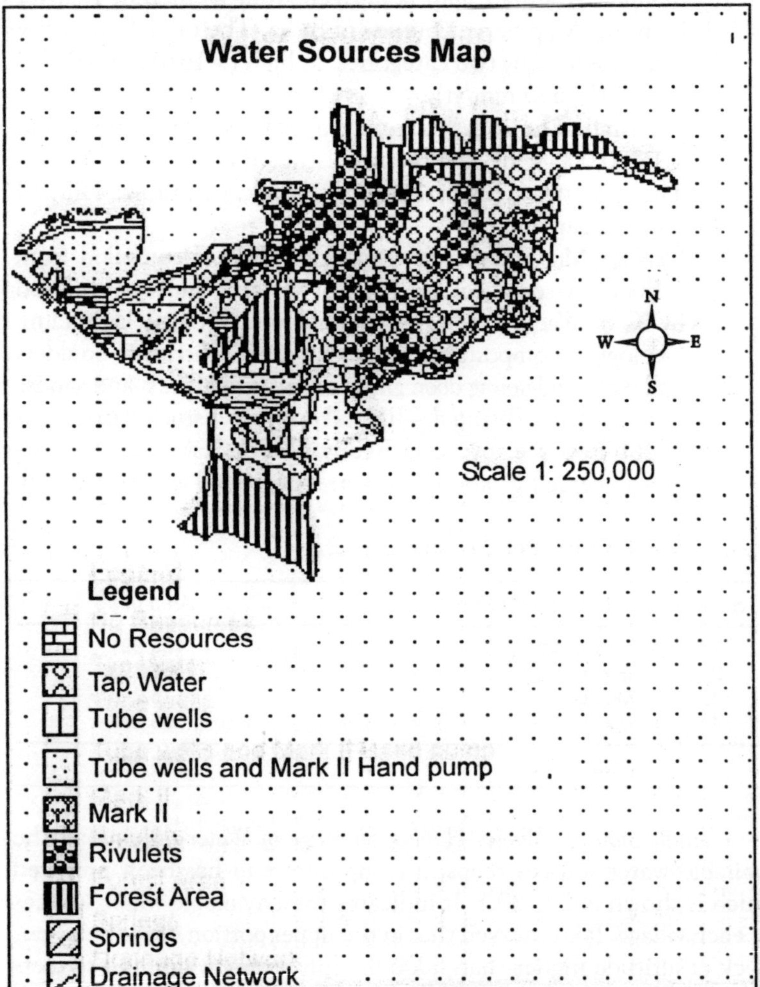

Figure 20.1: Water Sources Map

The composite map is overlaid with the population map to find out the shortage of water according to its demand for each village. The demand is calculated for the minimum requirement of 40 lpcd for drinking water. Supply of water is computed by the discharge of available water sources collected from Garhwal Jal Sansthan, Dehradun. Tube wells in the area are mainly used for irrigation purpose and their discharge is quite high. Only 10 per cent of water from tubewells is used for drinking purpose, i.e. average 200 lpm. India Mark II hand pump is designed for 12 lpm of 20 hrs. per day. Demand and supply of each village is compared through AML programming and shortage of water is computed for each village.

Table 20.2: Classification of Shortage of Water

Sl. No.	*Classification*	*Remark*
1.	Severe shortage	No sources of water present
2.	Shortage	Only piped water supply available (40% supply)
3.	Medium shortage	Demand greater than supply (> 40% supply)
4.	No shortage	Supply greater than demand

The 'shortage' of water map (Fig. 20.2) is overlaid on to the ground water potential map through AML programming to study both in combination. The villages having shortage of water according to the available water sources and ground water potential have been identified. Villages under the class of 'Severe shortage' category are found to have poor ground water potential. In these villages, only surface water and spring water can be used. Villages under 'shortage' category can use ground water for drinking purpose. Villages under 'medium shortage' contains with medium or good ground water potential, so ground water can be used for the drinking water supply.

Results and Discussions

1. Two villages, i.e. Bhagwanpur and Dungal gaon have been found under 'severe shortage' category, where water availability is minimum in the entire block. It has been observed that village Birsani has sufficient quantity of water through rivulets, as it is situated at higher elevation and the slope direction is also towards village Bhagwanpur. It is suggested that water

may be collected in the reservoir and brought down to the village under gravity. In case of emergency, a small quantity of water can also be pumped only during monsoon period. In both the villages Bhagwanpur and Dungal gaon, supply of water under gravity is proposed. The local population need to be educated for optimal water consumption. The used water may be utilized by disposing in land disposal system.

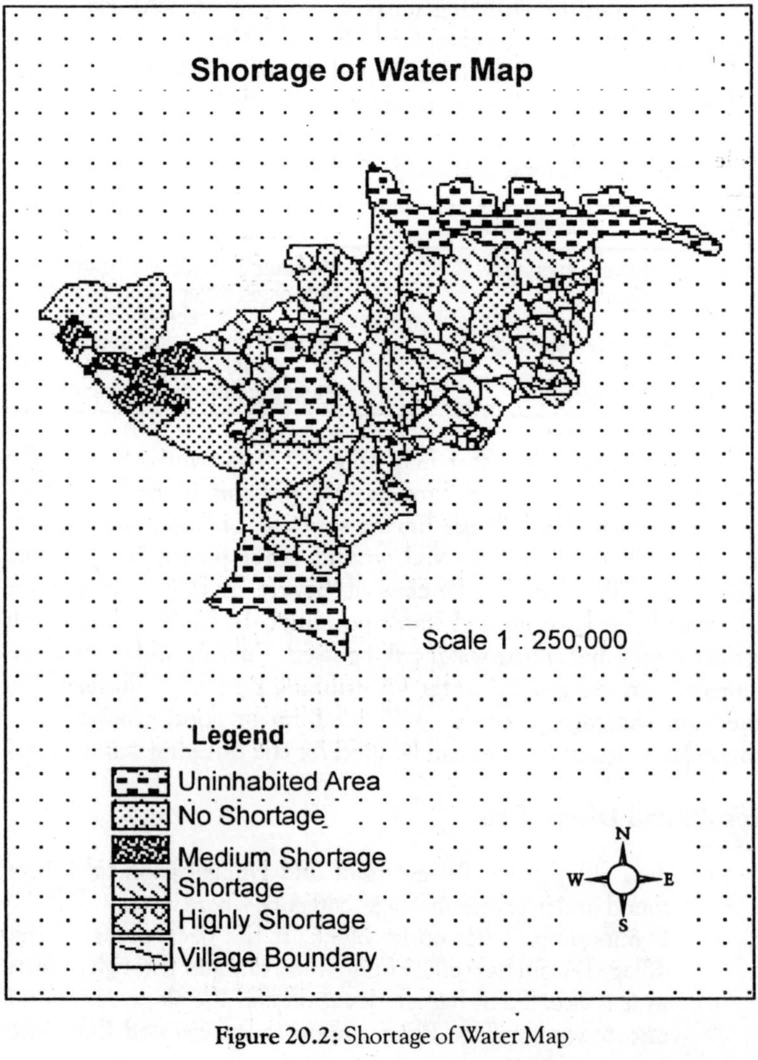

Figure 20.2: Shortage of Water Map

2. Dungal gaon is situated at high altitude in the hilly area. Ground water potential of this village is under 'Good' category, but due to geological condition and soil type, ground water withdrawal seems to be very difficult. In village Makreti, spring water is used for drinking purpose. It is available in plenty, and hence water from the spring can be stored in a reservoir in this village for water supply in dry period. Population of this village is only 94.
3. Water defecit villages are those where only 40 per cent supply is available through piped water supply and other private sources. Villages having shortage and good ground water potential are situated at high altitudes, with steep slopes and Blaini-krol geology. So, there may be difficulties in withdrawing the ground water. In these villages, rainfall is also high, so rainwater harvesting is suggested. Rainwater harvesting scheme utilizes the natural gradient which is easily available in hilly areas. The rainwater may be harvested either direct trap or roof catchment or ground catchment for a larger community. Because of hilly terrain, it is difficult to have sink wells, and hand pumps, therefore, rain water harvesting is suggested. In hilly areas, each hamlet, and not the whole village, will have to be taken as a single unit.

   In villages, where environmental conditions are suitable for ground water withdraw, ground water potential is found to be either poor or medium. There are 22 villages where ground water is poor, and if ground water is used in large quantity, strata is shallow at places. Ground water can be used in emergency only. Rainwater harvesting can be done for recharging of ground water to arrest the declining trend of water table.
4. Villages under 'medium shortage' are Sahaspur, Shankarpur, Indripur, Khushlpur. These villages have piped water supply, tube-wells and hand-pumps, but in spite of all these, water supply is inadequate because of large population of the area. Also piped water supply is available only for few hours, which creates unsanitary conditions for the poor rural population. In these areas, the ground water is very good as the rain water flows down from both the sides of the catchment which are higher in altitude. Soil of this area is bhabhar soil, and geology is young doon gravel, which is main aquifer used for the water supply. It is recharged through infiltration from surface and

alluvial fans. In dry periods, it is recharged by river water. So, it is proposed that Mark II hand pumps may be used in these areas.

Conclusions

1. Bhagwanpur and Dungal gaon are found under severe problem of water scarcity, as no water source is available in these villages. Ground water withdrawal is also difficult in these villages due to their geology, slope and other conditions. So, surface water available in immediate surrounding villages at the higher elevation source of water for these villages.
2. Villages situated just below the Mussoorie hills, have shortage of drinking water. The mountain range is rocky, and therefore, ground water potential tapping would not be economical. In these villages only pipe water supply exists presently, which is not sufficient. Hence these villages may be used for rainwater harvesting, e.g. roof top rain harvesting.
3. Villages having medium shortage of water have very good ground water potential. More field based study, however, is needed to decide the optimal use of water in these villages.
4. GIS has been found to be a useful tool for overlaying various thematic layers and integrating them to derive useful new maps, efficiently and accurately.

REFERENCES

Aronoff, S. (1989). *Geographical Information Systems: A Management Perspective*, WDL Publications, Ottawa (http://geog.hkbu.edu.hk/QZone/Teaching/GEOG3142/lect-01.html).

Bobba, A. G. (2001). Sustainable Development of Drinking Water in Indian Cities by Eco-Watershed Management Approach, *Hydrology Journal 24*, Vol. 2, pp. 1-20.

Chourasia and Gour, H. S. (1992). *Application of GIS and Remote Sensing for Identification of Artificial Recharge Sites and Groundwater Exploration*, India, http://www.cig.enmp.fr/~iahs/maastrich/w5/w505.htm.

Draft Perspective Plan for Dehradun District, (1992-2001). Vol. III, District Planning Unit Lalbahadur Shastri National Academy of Administration, Mussoorie.

Garg, P. K. (2000). "Appropriate Technologies for Surveying and Mapping the Features Related to Drinking Water," International Conference on Innovative Technologies for Rural Water Supply and Environment Sanitation, Roorkee, October 12-13, pp. 263-267.

Garg, S. K. (1992). *Water Supply Engineering*, Khanna Publishers, Delhi.

Maudgal, S. (1999). "Challenges in Sustainable Development Planning, Proceedings of National Workshop on Challenges in the Management of Water Resources and Environment in the Next Millennium: Need for Inter-Institute Collaboration", October 8-9, 1999, Dept. of Civil Engg., Delhi College of Engg., Delhi, pp. 139-145.

NEERI, (1994). "Carrying Capacity Based Developmental Planning of Doon Valley", National Environmental Engineering Institute, Nagpur.

Peavy, H. S., Rowe D. R., and Tchobanoglous, G. (1991). *Environmental Engineering*, McGraw-Hill Book Company, London.

Planning Atlas, (1992). Part B, District Dehradun, District Planning Unit, Lal Bahadur Shastri National Academy of Administration, Mussoorie.

Index